WILDLIFE CONTRACEPTION

ZOO AND AQUARIUM BIOLOGY AND CONSERVATION SERIES

SERIES EDITOR
Michael Hutchins, *Georgia Institute of Technology*

Published in cooperation with the American Zoo and Aquarium Association

This series publishes innovative works in the field of zoo and aquarium biology, conservation, and philosophy. Books in the series cover a wide range of relevant topics, including, but not limited to, zoo- and aquarium-based field conservation, animal management science, public education, philosophy, and ethics. Volumes range from conceptual books such as *Ethics on the Ark: Zoos, Animal Welfare, and Wildlife Conservation* to taxon-specific titles such as *Komodo Dragons: Biology and Conservation*.

WILDLIFE CONTRACEPTION

Issues, Methods, and Applications

Edited by Cheryl S. Asa and Ingrid J. Porton

The Johns Hopkins University Press
Baltimore

 Published 2005
9 8 7 6 5 4 3 2 1

The Johns Hopkins University Press
2715 North Charles Street
Baltimore, Maryland 21218-4363
www.press.jhu.edu

Library of Congress Cataloging-in-Publication Data

Wildlife contraception : issues, methods, and application / edited by Cheryl S. Asa and Ingrid J. Porton.

p. cm. — (Zoo and aquarium biology and conservation series)

Includes bibliographical references and index.

ISBN 0-8018-8304-0 (hardcover : alk. paper)

1. Animal contraception. I. Asa, Cheryl S. II. Porton, Ingrid J. III. Series.

SK356.C65W56 2005

639.9Ǝ3—dc22 2005010354

A catalog record for this book is available from the British Library.

This book is dedicated to the memory of Dr. Ulysses S. Seal, a man of great vision, who was undeniably the father of contraceptive use in captive animals, and who also played an important role in early efforts to apply contraception to free-ranging wildlife. He introduced the MGA (melengestrol acetate) implant to North American zoos in the mid-1970s, thus transforming captive animal management, and he organized the first international symposium on wildlife contraception in 1988. Ulie inspired and encouraged us, counseled and mentored us. We are grateful to have known him and to have worked with him.

CONTENTS

M. HUTCHINS

FOREWORD

Ensuring a future for life on Earth is rapidly becoming a Herculean challenge. Employing its superior technology, one species—*Homo sapiens*—has, in a relatively short period of time, risen to dominate virtually every ecosystem on the planet, ranging from tropical rainforests to prairie grasslands to coastal marshes and oceans. In converting these once natural settings for their own use, humans have forced wildlife into relatively small islands of habitat surrounded by a sea of humanity. This progressive destruction and fragmentation of the natural world, combined with other anthropogenic influences such as environmental pollution, overhunting, and the introduction of nonnative species, is changing entire ecosystems. As a result, many species indigenous to these systems are now threatened with irreversible extinction. Furthermore, because species sharing the same habitats are often highly interdependent, the loss of one or more key species can result in a domino effect, eventually threatening many other species and the ecosystem as a whole. Human influences on the natural world are now so pervasive that unprecedented levels of human intervention will become necessary to restore habitats and conserve endangered fauna and flora. In an effort to reverse these trends, we must begin to build a toolbox filled with technologies that can assist us in preserving biological diversity. One invaluable component of this evolving toolkit is the relatively new field of wildlife contraception.

Contraception is the deliberate prevention of conception or impregnation. The term is usually used to refer to prevention of pregnancy in mammals, including humans. There are many possible approaches to this biotechnological challenge. Mammals gestate their young (that is, carry and nourish their developing offspring) inside the maternal body. Pregnancy results from internal fertilization, so the simplest methods of contraception are to keep males and females apart, to prevent sperm and eggs from making direct physical contact (for example, condoms or diaphragms), or to use surgical procedures intended to block the movement of gametes to their intended destinations (for example, vasectomies or tying the fallopian tubes). The same effect is obtained when the ovaries or uterus are surgically removed (that is, spaying); the male equivalent of this procedure is castration (that is, removal of the testes). If fertilization does occur, the zygotes are transported to the uterus, where they begin development after becoming implanted in the uterine wall. Some contraceptive methods prevent ovulation by suppressing ovarian function (for example, subdermal hormone implants, oral contraceptives) and implantation can also be prevented (for example, intrauterine devices, IUDs). Immunization theory has recently led to several promising contraceptive methods. These techniques involve using vaccines to trick the body into rejecting certain molecules associated with reproduction. Thus, immune responses may be used to interfere with sperm function, female hormonal cycles, or fertilization.

Each of these existing contraceptive techniques has associated costs and benefits. Some methods are more effective at preventing conception or pregnancy than others; some are applicable specifically to only one sex; some have serious side effects, whereas others appear to pose fewer health risks; some are very expensive to produce, while others are less costly; some work better on certain taxa than they do on others; and some methods, although practical for humans or even for zoo or aquarium animals, cannot be effectively applied in free-ranging animals. For example, many hormone-based techniques involve daily applications or treatments, which would be impractical for use with wild animals. In addition, wild animals may have social behaviors that reduce the effectiveness of certain techniques. In certain polygamous species, for instance, sterilizing socially dominant males may have little effect on reproduction, especially if females also mate with subordinate males (that is, so-called sneaky copulators).

Contraception is one manifestation of the broader concept of reproductive control. Thus, *wildlife contraception* is the prevention of conception or pregnancy in wild (that is, nondomesticated) animals. Typically, we think of wild animals as animals that are free living, that is, living their lives independent of humans. However, some wild animals also live in human care. Thus, the term wildlife contraception is also commonly applied to contraception in wild (nondomesticated) species held in zoological parks or aquariums.

The goal of wildlife contraception is to control reproduction in individuals, but ultimately it is also to reduce or stabilize the size of animal populations. However, contraception is just one of many possible approaches to population control. By preventing fertilization of the ovum or implantation of the fertilized egg in the uterus, reproduction is controlled at its earliest stages, but control can be and is applied at later stages as well. Other options include removing the embryo or fetus during development (abortion), killing infants shortly after birth (infanticide), killing juveniles or adults (often referred to as culling), and translocation (moving animals from one location to another). Each option carries its own practical limitations and ethical concerns.

The discussion here has focused on mammals, because many animal species do not become pregnant, but rather deposit their fertilized eggs into the environment (for example, most amphibians and reptiles). Thus, in these species, it is often easier to simply prevent the development of fertilized eggs as they are laid (that is, to practice ovicide).

There is a wide range of opinions regarding when and how wildlife populations should be controlled. However, as a general rule, ethical considerations become exceedingly more difficult and complex when population control involves killing healthy, sentient individuals at later stages of life. Intervention before fertilization or implantation seems to be preferable from an ethical point of view, as it is hard to imagine sperm, eggs, or even early-stage embryos as being capable of experiencing pain or distress. This perception explains, in part, why many humans have had such a difficult time deciding where to draw the line with abortion (that is, at what age is the fetus considered to be a "person?"). It also explains why contraception has been embraced by animal protection organizations as a method of controlling wildlife populations. It is important to note, however, that contraception, although considered by many persons to be a more palatable method of controlling wildlife populations than killing individuals, has many disadvantages. For example, in the case of long-lived species, even should contraception be successful, many years must pass before overall population numbers are reduced. Thus, the problem could conceivably continue until the population is eventually reduced through natural attrition (that is, through normal mortality). In addition, some hormone-based contraception techniques are known, in some species but not in others, to have deleterious side effects such as cancer and other health-related concerns. Some contraceptive techniques may result in permanent sterilization or alter normal behavior. These effects can also limit their application under certain circumstances, such as in zoo or aquarium populations where reversibility is often desired. For example, it may be desirable to prevent inbreeding if sexually mature animals are temporarily housed with close relatives. Contraception could then be reversed when more appropriate mates (that is, unrelated animals) become available.

It is easy to understand why humans need to control populations of domestic animals, such as farm animals or pets. These animals are completely dependent on human care. When their populations become too large, our ability to provide appropriate care can be compromised. In turn, this can lead to suffering and loss of life. Other problems can ensue, such as increased cost, overgrazing of pastures, or a higher incidence of disease in the case of overcrowded domestic livestock. Overpopulation of pets, such as domestic dogs and cats, is also problematic. More than 27 million dogs and cats are impounded in shelters annually in the United States and some 17 million of these must be euthanized. This is a complex problem, from both an economic and an ethical perspective. Much of the blame should be attributed to irresponsible owners who fail to prevent their pets from producing unwanted young. Therefore, population control is essential if we are to ensure that all existing dogs and cats can be placed in good homes. In addition, overpopulation increases the chances that neglected animals will have to fend for themselves. Free-roaming (feral) domestic dogs and cats take a tremendous toll on wildlife populations and can also spread dangerous diseases, such as rabies and toxoplasmosis, to humans and other animals.

It is perhaps more difficult to explain why it is becoming increasingly necessary to control populations of wild animals. It seems paradoxical to need to prevent the reproduction of free-ranging wildlife, especially when so many species are declining precipitously. However, there are many cases in our human-dominated world in which wildlife populations should or even must be controlled. Indeed, control is sometimes necessary to protect human health or life, domestic livestock, or crops, to preserve endangered fauna, flora, and habitats, and to maintain public support for national parks and equivalent reserves. Three examples of the complexities involved in wildlife contraception and population control are provided here for purposes of illustration.

HUMAN–ELEPHANT CONFLICTS

African and Asian elephants (*Loxodonta africana* and *Elephas maximus*) are the largest of extant land mammals. Both species have suffered greatly at the hands of humans. Tens of thousands of elephants have been killed for their valuable ivory and meat. The end result is that elephant populations have spiraled downward during the past few decades. Both Asian and African elephants are now classified as endangered by IUCN–the World Conservation Union.

Among the most significant threats to the elephant's continued existence is the potential for direct conflict with humans. Growing human populations in Africa and Asia have gradually encroached on elephant habitat, and elephants and people now live in close proximity on both continents. Elephants frequently raid agricultural crops, and hundreds of people are killed annually attempting to protect their livelihoods from these large, potentially dangerous herbivores. Many elephant range countries are developing economies, so food crops are critical for human health and survival. These problems are exacerbated when elephants become locally overabundant and population pressures force them to leave the confines of national parks or equivalent reserves. Densely packed elephant populations also can cause substantial damage to the ecosystems in which they live. In particular, hungry elephants are known to tear down or damage vast numbers of trees, thereby converting woodlands to grasslands or semidesert—an activity that can have deleterious impacts on numerous sympatric species. Progressive habitat degradation can eventually have a negative impact on elephants as well. As food availability decreases, competition increases, eventually resulting in widespread starvation, suffering, and death.

The traditional response to local elephant overpopulation has been to reduce numbers through culling. This practice involves shooting of elephants by park rangers or contracted professional hunters. Generally, entire family groups, including infants, juveniles, and adults, are killed almost simultaneously. Elephant family groups, which consist of related adult females and their young, are extremely cohesive, and the thinking behind this approach is that it is preferable to destroy all family members at once rather than to separate mothers from their young. Meat from culled elephants is often given to and eaten by hungry local people.

Negative human–elephant interactions raise complex and excruciatingly difficult animal management questions, the answers to which are going to be driven by current ecological, political, and socioeconomic realities. Elephant culling may be preferable to slow death from starvation and may also help to prevent deaths of local people by reducing the potential for human–elephant conflicts. However, some animal protectionists, including People for the Ethical Treatment of Animals (PETA) and the Humane Society of the United States (HSUS), have vehemently criticized such practices as being cruel and inhumane. Anyone who has witnessed an elephant culling—whether on film or in person—can sympathize with this point of view, which explains why animal protection organizations have embraced wildlife contraception and translocation as the preferred methods to control elephant populations. Recently, HSUS and the International Federation of Animal Welfare (IFAW) paid the South African government millions of dollars to stop elephant culling temporarily. They have also put pressure on other African countries to cease culling activities. In addition, these organizations have aggressively funded experimental tests of contraceptive vaccines to assess their effectiveness in slowing or reversing population growth in elephants.

Animal protectionists focus on the welfare or "rights" of individual animals, whereas conservationists are more concerned with ensuring a future for populations, species, and ecosystems. Sometimes these two worldviews come into direct conflict. Being a nonlethal alternative, contraception would likely be the most preferable option for both camps, except for its obvious limitations. The most significant drawback of elephant contraception is its inability to impact the population in the short term. Elephants can live up to 60 years or more and, even if an effective contraceptive technique could be found, it could take many years before populations begin to decline through normal attrition. In the meantime, the problems associated with elephant local overpopulation are expected to continue to escalate. Many national parks in Africa are displaying the consequences of inaction, with widespread destruction of woodland habitats occurring at an alarming rate. Some previously lush areas now resemble war zones. In addition, hundreds of people continue to be killed and numerous crops are destroyed annually, potentially eroding local and national support for protected areas and wildlife conservation.

CONTROLLING INVASIVE SPECIES: BRUSH-TAILED POSSUMS IN NEW ZEALAND

Another potential application of wildlife contraception is to control or eliminate populations of nonnative or invasive species. Invasive species are those that are transported by humans—either purposefully or accidentally—across natural barriers to dispersal (for example, oceans, mountain ranges), where they may subsequently invade and become established in a host ecosystem. Nonnative invasive species have had many deleterious impacts throughout the world, and their control is often necessary to protect native fauna and flora, natural habitats, or human health.

One such example was the introduction of brush-tailed possums (*Trichosurus vulpecula*) to New Zealand. These rabbit-sized arboreal marsupials are native to Australia and were imported purposefully during the nineteenth century, apparently in an effort to establish a fur industry. Possums found New Zealand's habitats and climate so accommodating that their populations have soared to an estimated 60 million. The possum's success probably occurred because the island's forests and vegetation evolved in the absence of mammalian browsers and thus there was little competition for resources.

Brush-tailed possums have become serious wildlife pests in New Zealand, where they threaten the survival of native plants and animals and spread diseases. The activities of the introduced possums are reducing the survival and reproductive success of highly endangered endemic mammals and birds, such as the short-tailed fruit bat (*Mystacina tuberculata*) and kokako (*Callaeas cinerea*). The possums are also responsible for spreading bovine tuberculosis, which is a serious threat to the country's agricultural industry.

Recognizing the seriousness of the problem, the New Zealand government commissioned a study to evaluate various methods to control possum populations, including contraception. It was concluded that much more research is needed to understand the species reproductive biology and to assess the potential for contraception to make a dent in the massive population. With so many possums to eliminate, it is deemed critical to identify cost-effective solutions that were also acceptable to the public. It was further suggested that a combination of traditional control techniques and modern contraceptive technology offers the only chance of reversing the negative impacts of this major biological invasion.

Conservationists have recently identified introduced species as among the most significant threats to wildlife and nature. Once again, it is important to note that some animal protectionists, with their focus on individual animal welfare and "rights," have vehemently opposed efforts to control populations of invasive species, especially when such control involves killing. This opposition remains intact even when population control is considered necessary to save native species from the threat of extinction. In fact, the father of animal rights, Tom Regan, has characterized any attempt to usurp the rights of individual animals to preserve endangered species or ecosystems as "environmental fascism." The use of wildlife contraception can help to reduce some of these philosophical conflicts, but, once again, it is questionable if any control technique can be called effective if it takes many years for the intended effects to be realized. This caveat is especially true when the negative impacts are expected to continue until the target population eventually becomes reduced through natural attrition.

MANAGING ANIMAL POPULATIONS IN ZOOS AND AQUARIUMS

Accredited zoos and aquariums are playing an increasingly important role in wildlife conservation. With an estimated 140 million visitors each year, North American zoological institutions are becoming important centers for conservation education and awareness building. Additionally, these institutions have become increasingly involved in wildlife research and field conservation, supporting some 2,200 projects in more than 90 countries annually in collaboration with governmental, nongovernmental, and university partners. However, of particular interest to these discussions are zoo and aquarium breeding programs.

Zoo and aquarium breeding programs have become highly organized. Populations of selected species are managed cooperatively and scientifically. Computerized records (collectively called studbooks) are maintained on each individual in the breeding population. These databases are analyzed and used to develop breeding master plans, the purpose of which is to maintain as much gene diversity in the population as possible. Genetic diversity is critical for sustaining relatively small, isolated populations over time, and animals are moved between institutions to avoid inbreeding, maintain rare alleles, and obtain the best possible matches.

Zoo and aquarium managers must also be concerned about population demographics; to sustain a population, it is important that sufficient numbers of animals are retained in each sex and age class, with larger numbers in the younger age classes and fewer in the older age classes (that is, a classic age-pyramid structure). Many of the species managed in zoo and aquarium breeding programs are threatened or endangered, and some are being used to reestablish extirpated wild populations (for example, black-footed ferrets, California condors). This concern makes it essential that populations are properly managed, and wildlife contraception has become an important tool for realizing this goal. One important application of wildlife contraception in zoos and aquariums is to curtail or prevent reproduction in genetically overrepresented individuals. Such individuals, if allowed to continue to reproduce, could swamp the captive population with their genes, thus resulting in the loss of important gene diversity.

Accredited zoos and aquariums are also concerned that captive wild animals receive appropriate, professional care. In the past, some institutions were criticized for sending animals to substandard facilities. However, modern zoos and aquariums include animal welfare among their highest priorities; this is why collections are now carefully planned and managed to ensure that populations do not exceed their available space. Production of animals that are "surplus" to managed populations might increase the chance that some individuals could end up in the hands of those not qualified to care for them. In addition, being public institutions, zoos and aquariums are unlikely to utilize less palatable methods, such as culling, as a routine method of population control. Consequently, wildlife contraception has become an increasingly important tool for zoo and aquarium managers.

Fortunately, zoo and aquarium personnel do not share the same limitations as wildlife managers, thus making it possible to apply a wider variety of contraceptive techniques. For example, some zoo animals, such as great apes, can be trained to take oral contraceptives daily, whereas pills would be an unlikely choice for use on free-ranging animals. Hormonal implants (placed under the skin) are very practical for use in zoos and aquariums and are a popular contraceptive method for hoofstock, such as antelope or gazelles, and large carnivores, such as lions or tigers. Because many animals are utilized for breeding, reversibility of contraception is often an important requirement.

The wide diversity of species displayed and bred in zoos and aquariums (15,000 worldwide) presents a significant challenge for contraceptive researchers. Some contraceptive methods are effective and safe for some taxa but not practical, safe, or effective for others. It is therefore important that zoological managers be kept informed about the latest research developments. Zoos and aquariums and their animal collections are becoming important venues for testing new and innovative contraceptive techniques that can then be applied to free-ranging animals.

CONCLUSIONS

Wildlife conservation and management are becoming exceedingly complex and difficult human endeavors from both a practical and an ethical perspective. That difficulty is precisely why this book and others that will follow it are so important. There is certainly a need to research and develop new and effective methods of human, domestic animal, and wildlife contraception. There is also a need to continue to debate the reasons to control wildlife populations and the circumstances under which various techniques can and should be applied.

Public opinion and politics will undoubtedly play a role in identifying solutions, but it is essential that key decision makers and interested laypersons understand all the options and implications before deciding on a potential course of action or inaction. Solutions are particularly difficult when different value systems lead to conflicting perspectives. Wildlife contraception has the potential to reduce conflicts between those who believe that wildlife should be "left alone" and those who believe that some wildlife populations should and, in fact, must be managed.

Wildlife contraception science, of which many aspects are on display in this volume, represents a wealth of innovative solutions to difficult issues in conservation biology. Scientists who specialize in this area of applied biology should be applauded and encouraged to expand their important and useful work. Until humans learn to effectively control their own populations and reduce their impact on the natural environment, it is necessary to actively intervene and maintain the tenuous balance between wildlife and humans and between wild animals and their heavily impacted environments.

PREFACE

The first time we realized there was a dire need for accurate and shared information regarding contraceptive methods applicable to zoo animals was in 1989 when one of us (I. J. P.) participated in a Lion-Tailed Macaque Species Survival Plan Masterplan meeting. The meeting produced a set of recommendations for a number of individuals to breed, but the greater challenge was presented by the recommendations for others to *not* breed. The latter concept evoked discussion and concern because, although the melengestrol acetate (MGA) implant was a contraceptive method used in zoo animals, the meeting participants had heard various reports that the implants were not effective in some primates. Stories such as one in which a female orangutan was treated with an MGA implant but gave birth a year later raised worried questions. If MGA implants were not effective, what contraceptive methods could be used to control reproduction in a species such as the lion-tailed macaque?

The issues raised at the Masterplan meeting led us to look into what information was actually available to animal managers on the efficacy of contraceptive methods in exotic animals. We realized there was no established venue through which zoos could share their experience with zoo animal contraception. Subsequent discussions with American Zoo and Aquarium Association (AZA) staff and the AZA Wildlife Conservation Management Committee led to the proposal for a fact-finding committee to help the zoo community with this issue. That committee was designated a Task Force with the expectation that, when the relevant information was gathered, the task would be completed and the group disbanded. Instead, the AZA Contraception Task Force evolved into the AZA Contraception

Advisory Group (CAG), and 15 years later the need for information and research remains as great now as it was then.

Members of the Contraception Task Force included a range of zoo professionals who could identify and address the challenges inherent in exotic animal contraception, including reproductive biologists, veterinarians, curators, and a pathologist. The first objective of the group was to ascertain what contraceptive methods were actually being used in zoo mammals, and from that approach the AZA Contraception Survey was conceived. In addition, Task Force members also monitored contraceptive research that was applicable to zoo animals. In turn, that step led to recommending and coordinating research on existing and potential contraceptive methods for exotic mammals within the zoo community.

The first Contraception Task Force Survey, sent to AZA institutions in the fall of 1989, was for primates only. The Survey was divided into two parts, one questionnaire for MGA implants and a second questionnaire for all other contraceptive methods. In the summer of 1990 a second Survey, for carnivores, was distributed. Today the Survey requests data for all mammalian species and is distributed to other national and international zoological facilities as well as those of the AZA. Information from the Surveys is entered in the Contraception Database, combining the experience of all the contributing institutions into one file. The Contraception Database is currently composed of more than 18,000 records but, as can be seen from the chapters in Part III, information on the efficacy and reversibility of the various contraceptive methods for the variety of mammals that we treat is slow to accumulate.

To provide the zoo community with guidance in the selection of appropriate contraceptive methods for zoo animals, the Contraception Advisory Group developed the "Contraception Recommendations." The Recommendations, based on survey results and research findings, provide taxon-by-taxon guidelines regarding the known efficacy and safety of available contraceptive methods. The Recommendations are updated annually and are available to animal managers through our Web site: www.stlzoo.org/contraception.

In addition to monitoring contraceptive use, the Contraception Advisory Group has coordinated research efforts to develop and test promising new methods. We looked, in particular, for male-directed methods, to provide a wider selection to managers, and for nonsteroidal methods for carnivores, in which we were seeing side effects from progestin use. In a collaborative study that included 15 zoos, we evaluated injectable vas plugs in a wide variety of species. Although sperm passage could be effectively blocked, efficacy was associated with sufficient damage to interfere with reversal (L. Zaneveld and C. Asa, unpublished data). Male domestic cats were used in a study of an antispermatogenic agent, an analogue of indenopyridine. Unfortunately, the project was abandoned after diarrhea, sometimes containing blood, was observed the day after treatment (P. Fail, D. Kunze, and C. Asa, unpublished data). Even more frustrating, however, was our experience with bisdiamine, another antispermatogenic compound. Despite a demonstration of efficacy in a study with gray wolves (*Canis lupus*), we could not locate a company that could produce the drug at an affordable price.

More recently, we have been more fortunate and have been able to introduce a new product, MGA incorporated in an herbivore diet, in cooperation with Purina Mills, LLC. In addition, we have contracted with Peptech Animal Health in Australia to import their new product, Suprelorin, a gonadotropin-releasing hormone (GnRH) agonist. We will continue to look for promising new options and new partners to provide more contraceptive alternatives to the zoo and wildlife communities.

As the contraception program grew in scope and complexity, AZA recognized the need to centralize its efforts. Thus, in 1999 AZA established the Wildlife Contraception Center at the Saint Louis Zoo, in recognition of the Zoo's longstanding commitment to the development and implementation of wildlife contraception. Continuing as advisors to the center are the members of the Contraception Advisory Group, which still includes reproductive physiologists, curators, and veterinarians but has expanded to incorporate a number of scientific and clinical advisors knowledgeable in contraception, all dedicated to providing the resources for better contraceptive management.

ACKNOWLEDGMENTS

We thank the Saint Louis Zoo, and in particular Charlie Hoessle, for allowing us to devote our time to building and running the contraception program and for support of program activities, and Dr. Jeffrey Bonner for allowing us to continue this work. And, of course, we are indebted to Dr. Ulysses Seal for starting it all and to Dr. Ed Plotka for picking up the reins and continuing to supply melengestrol acetate (MGA) implants to the zoo community and MGA reports to the Database. We are grateful for the design and programming efforts of Betsy Hornbeck, Pablo Molina, and Karen DeMatteo, because without them the Contraception Database would not exist. We also offer our thanks to Cynthia Holter, Chris Cannon, and Jeff Huntington of the Saint Louis Zoo Development Department for helping raise funds; to the Geraldine R. Dodge Foundation and the Barbour Foundation for program support; to Janet Powell for managing the AZA Wildlife Contraception Center Web site; to Kevin Mills for the AZA Wildlife Contraception Center logo and brochure design; to Dr. Joan Bauman for hormone analysis for contraceptive monitoring; and to Jan Dempsey for nutrition management of the MGA feed program. Last, but perhaps most important, we are immensely grateful to the dedicated people at zoos around the world who complete and submit the surveys that make up the Contraception Database and to the zoos that have participated in research trials of promising new contraceptive methods. This book would not have been possible without them. The summaries, analyses, and recommendations in these chapters are based on the data generated through their cooperation.

CHERYL S. ASA AND INGRID J. PORTON

INTRODUCTION
THE NEED FOR WILDLIFE CONTRACEPTION
Problems Related to Unrestricted Population Growth

The balance of nature is a long-standing popular concept of the interactions of free-ranging wildlife populations, exemplified by the 10-year cycle of the Canada lynx *Felis lynx* and the snowshoe hare *Lepus americanus* (Elton and Nicholson 1942). In ecosystems undisturbed by humans, predators and prey, and forager and forage, have coevolved. Wildlife populations are dynamic, but even the extreme shifts in numbers described for the lynx and hare do not result in species extinctions, although local populations may crash as a result of local conditions.

Prey species are obviously impacted by predation, but even predators seem to have evolved other intrinsic mechanisms that help keep their numbers in balance with local conditions. Although best studied in rodents and ungulates, reproductive rates in most species are limited by food availability. Food restriction has been shown to reduce the levels of reproductive hormones, and thus curtail reproductive processes, in both male and female mammals (Piacsek 1987; Bronson 1989). Even more impressive are the density-dependent mechanisms described for some rodents, in which crowding, even in the presence of adequate food, can suppress reproduction via a stress response mediated by the adrenal glands (house mice: Christian and Davis 1964; deer mice: Terman 1973). Of course, food limitation and drought can have a more profound effect on population numbers through starvation and dehydration.

So, if animals are indeed adapted, at least to some extent, to control their own numbers, why might there be a need for fertility control for free-ranging wildlife? The conflict comes at the human–animal interface (Jewell and Holt 1981). Historically, humans have altered wildlife abundance and distribution in many ways. To protect themselves and their livestock, humans have eliminated or reduced the number of predators such as wolves and foxes. Yet despite the centuries humans have done battle with coyotes (*Canis latrans*), the numbers of this wily canid have not declined. In fact, coyotes are more widely distributed in the United States now than ever, probably because of the much more successful program to exterminate wolves, as wolves seem able to outcompete and to kill coyotes wherever their ranges overlap.

The elimination of predators such as wolves throughout most of the United States simplified efforts by state fish and game agencies to manage for the benefit of white-tailed deer (*Odocoileus virginianus*) as a game species. Those programs have been so successful that deer are now considered pests in many parts of the United States. The deer population problem has been exacerbated by human population growth and spread. Not only are people increasing in numbers, but their per capita use of resources is multiplying even more rapidly (Liu et al. 2003). Urban sprawl into the surrounding countryside, especially in the United States, brings people directly into contact with wildlife. Some residents enjoy sharing space with deer and coyotes, but others resent the encroachment of deer into their living space, particularly when the deer consume ornamental plants and trample gardens. A more serious threat is the increasing likelihood of vehicle collisions with deer.

Other ways that humans influence animal distribution include the attraction of rodents to agricultural fields and storage facilities, which then attracts snakes. Livestock that during domestication have lost the natural predator avoidance behaviors of their ancestors provide easy meals for wolves, coyotes, foxes, and other carnivores. In and around cities, people feed pigeons and Canada geese (*Branta canadensis*), creating much larger than natural local populations that increase the chances of spreading disease, not to mention the increased amount of wastes.

Humans have introduced, purposefully or unwittingly, both nonnative and domestic species to new habitats and new continents. Introduced species such as European rabbits (*Oryctolagus cuniculus*) and red foxes (*Vulpes vulpes*) have been especially disastrous for Australian native wildlife and ecosystems (Wilson et al. 1992). Feral domestic species such as pigs and goats also have wreaked havoc on local ecology. But perhaps the most emotion and money have been spent on attempts to control or protect populations of feral horses in the United States (Asa 1992; Symanski 1996). In the middle of the twentieth century, when the tractor replaced the workhorse, ranchers released their horses onto rangeland. Today, federal laws prohibit killing or removing the feral horses that are their descendants, and in the absence of wolves and with few puma as predators, the horse populations have been documented to increase at the rate of 15 to 20 percent per year (Eberhardt et al. 1982; Wolfe 1986).

The population imbalances that have resulted from prior human disturbance or interference now require informed management by humans to prevent habitat destruction from such causes as overpopulation and overgrazing or extinction of other species resulting from competition or predation. Lethal control is no longer widely acceptable in the United States and much of Europe, and translocation, first proposed as an alternative to lethal methods, probably results in the death of many of the translocated animals themselves or of other animals in their new habitat through either direct or indirect competition. Few habitats are not already at carrying capacity, which means that newly introduced individuals are unlikely to find adequate resources or territory to survive without displacing other individuals. Add to this the threat of disease or parasite transmission, plus the stress of capture and transportation, and it becomes more evident why translocation is not advisable except in special situations.

Thus, contraception has become an attractive option for population control. However, despite decades of attempts, very few methods have been successfully applied to more than isolated populations. The difficulties with application include delivery, species specificity, and possible effects on humans (for example, deer entering the human food chain). Chapter 13 (this volume) summarizes the successes and the challenges.

Although the population dynamics of free-ranging wildlife differ considerably from those of their captive cousins, the problems pursuant to unrestricted reproduction are similar in both populations. Such problems in both cases relate to carrying capacity and resource availability and are the products of human manipulation. The rationale for managing reproduction in zoos is clearer, because humans bear full responsibility for the welfare of captive animals, whereas there is more disagreement about the appropriateness of interference with free-ranging populations.

Unrestricted reproduction became a problem in zoos when expertise in managing the medical, nutritional, and environmental needs of exotic mammals improved (Kisling 2001; Schaffer 2001). Infant survival increased, adult mortality decreased, and the predicament of insufficient space within zoos to house the animals that were produced came to the forefront (Conway 1976; Xanten 2001).

Originally, zoos were consumers of wildlife. Mammals as diverse as elephants, primates, big cats, and antelope were brought in from the wild by animal collectors and dealers. Animals died during capture, during transport, and after arriving at a zoo. There was no problem with surplus animals; rather, the priority issue was to replace the animals that died with additional individuals from the wild (Mench and Kreger 1996; Whittaker 2001; Hanson 2002). By the 1950s and 1960s, however, enough expertise had been gained in the management of exotic mammals that reproduction was more commonly achieved. Those zoos that were successful at breeding animals were able to supply the offspring to other zoos. Reproduction, especially in some species that bred well in captivity, such as lions, certain primates, and antelope, caused a surplus of offspring. Zoos sold the animals, and the derived income became an important component of the zoo's budget. Zoos sold animals not only to each other but also to dealers, and in most cases the final destination of the animals was unknown (Graham 1987; Xanten 2001). Dealers sold the animals to other legitimate zoos, but also to private breeders, ranchers, the entertainment industry, and into the pet trade. Animal dealers and zoos alike sold primates to laboratories for biomedical research. For several decades, finding facilities to accept surplus exotic animals was not a problem faced by zoos.

Changes in some animal transfer practices occurred with the adoption of the Convention on International Trade in Endangered Species of Fauna and Flora (CITES 1973) and the 1973 passage of the US Endangered Species Act (Xanten 2001). The environmental movement, epitomized by Earth Day, the animal welfare movement, and the increasing importance and influence of regional zoo associations, further changed animal exchange procedures. Progressively, reputable zoos became more concerned with wildlife conservation and upholding animal husbandry standards, which led to a decreased use of animal dealers to move surplus stock (Graham 1987). The International Species Information System (ISIS), developed in 1973, allowed zoos to better manage and share individual animal records. By the mid-1970s, breeding loans became a more routine procedure in zoos (Walker 2001).

The 1980s marked a decade that dramatically changed how zoos managed captive animal populations. Milestone events included the formation of American Zoo and Aquarium Association (AZA) Species Survival Plans (SSP) as well as the increased emphasis placed on monitoring captive populations through regional and international studbooks. Programs equivalent to SSPs were initiated in other regions [for example, the European Endangered Species Program (EEP) in Europe]. Regional zoo associations developed acquisition and deacquisition policies that began to play a more prominent role in guiding the transfer of animals outside of member institutions (Butler 2001). A heightened interest in controlling and reducing reproduction was another change that occurred in the management of zoo populations. Research targeted at developing a safe and effective contraceptive method for lions and tigers in the mid-1970s produced the melengestrol acetate (MGA) implant (Seal et al. 1976). The use of the MGA implant expanded beyond felids to other taxa, including primates and ungulates.

The 1990s saw a significant expansion in the number of SSP and EEP programs and the creation of Taxon Advisory Groups (TAG) (Wiese and Hutchins 1994).

TAGs are charged with developing Regional Collection Plans (RCP) that provide recommendations regarding the selection of taxa that should be held and managed through cooperative programs by member zoos. One objective of the regional collection planning process is to define the role and function of each taxon recommended for a captive management program. RCPs are leading to a reduction in the total number of taxa that are displayed in accredited zoos. Additional guidelines concerning the transfer of SSP species (AZA 1976), development of more stringent animal caretaking standards (Hutchins 2001), and an active debate on the ethical responsibility zoos have toward all animals they produce (Norton et al. 1995) have significantly decreased the number of facilities to which accredited zoos can send their animals.

Today, accredited zoos are at carrying capacity. Consequently, even when reproduction is carefully controlled, as is the case in SSP populations, there are surplus animals (Conway 1976; Lacy 1991). The question of how to address the surplus issue has been debated for years among zoo professionals within the accredited zoo community. Is it more humane and ethical to euthanize surplus animals that cannot be placed in appropriate facilities (Graham 1996; McAlister 2001)? How long can animals that have had to be removed from their social group be housed in inadequate off-display housing (Lindburgh 1991)? Is it more humane to transfer these surplus animals to nonaccredited zoos? The AZA Code of Professional Ethics (AZA 1976) permits animal transfers to facilities that can "care for them properly." But what standards define this proper care (Maple et al. 1995; Wuichet and Morton 1995; Kirkwood 2001; Hutchins 2001)? Are there different standards for different taxa? The AZA Accession/De-accession Policy differentiated nonhuman primates from other taxa with the directive that AZA institutions could never place nonhuman primates into the pet trade (AZA 2000). Should animals transferred out of AZA zoos be permanently sterilized to prevent the possibility that their offspring could be placed in substandard facilities or into the pet trade? Should zoos build and support retirement facilities for older animals that are no longer needed for breeding (Lindburgh and Lindburgh 1995)? The broad range of opinions assures the debate will continue as the accredited zoo community struggles with developing science-based definitions of animal well-being and the husbandry standards that meet these definitions.

With the moral question regarding how accredited zoos should manage the current surplus animal problem yet unresolved, unlimited reproduction is plainly an indefensible concept. It can only lead to two outcomes. First, a larger proportion of healthy animals born in zoos will need to be prematurely euthanized. And second, a larger proportion of animals born in zoos will ultimately end up in substandard and even inhumane facilities. Although contraception cannot totally eliminate the surplus animal problem that zoos face (Lacy 1991, 1995), unrestricted reproduction of animals within zoos is morally unacceptable (see Chapter 1, The Ethics of Wildlife Contraception).

Although there is considerable overlap in the principles and methods of contraception use in captive and free-ranging wildlife, there are substantially more data on the application of contraceptives in captive animals. The contraception program in US zoos began in the mid-1970s when Dr. Ulysses Seal began supplying MGA implants to control reproduction in lions and tigers (Seal et al. 1976). His database, containing information on all the implants he provided to zoos for the next 15 years, has been incorporated into the AZA Contraception Database, which now collects information on all contraceptive methods used by accredited zoos in the United States and, increasingly, by zoos around the world. Thus, not only has the contraception program for captive animals been in existence longer than any control program for free-ranging species, it is significantly more extensive. Because so many more data are available for captive wildlife, the coverage in this volume cannot avoid reflecting that imbalance. Although there have been numerous small trials of many promising contraceptive methods either in or intended for application to free-ranging wildlife, few have met with enough success to be implemented (see Chapter 13, Contraception in Free-Ranging Wildlife). Nevertheless, we hope that the information provided here may be of help to those working toward the development of techniques for free-ranging populations.

REFERENCES

Asa, C. S. 1992. Clash of interests in population control of free-ranging animals. In *Proceedings, Joint Conference of the American Association of Zoo Veterinarians and American Association of Wildlife Veterinarians,* 28–30, November 15–19, 1992, Oakland, CA.

AZA (American Zoo and Aquarium Association). 1976. *Code of Professional Ethics.* http://www.aza.org/AboutAZA/CodeEthics/.

AZA (American Zoo and Aquarium Association). 2000. *AZA Accession/De-accession Policy.* http://www.aza.org/AboutAZA/ADPolicy/.

Bronson, F. H. 1989. *Mammalian Reproductive Biology.* Chicago: Chicago University Press.

Butler, S. J. 2001. American Zoo and Aquarium Association. In *Encyclopedia of the World's Zoos,* Vol. 1, ed. C. E. Bell, 23–27. Chicago: Fitzroy Dearborn.

Christian, J. J., and D. E. Davis. 1964. Endocrines, behavior and population. *Science* 145:1550–1560.

CITES (Convention on International Trade in Endangered Species of Fauna and Flora). 1973. *Signed at Washington, D.C., on March 3, 1973. Amended at Bonn, on June 22, 1979.* http://www.cites.org/eng/disc/text.shtml.

Conway, W. G. 1976. The surplus animal problem. In *AAZPA Annual Conference Proceedings*, 20–24. Wheeling, WV: American Association of Zoological Parks and Aquariums.

Eberhardt, L. L., A. K. Majorowicz, and J. A. Wilcox. 1982. Apparent rates of increase for two feral horse herds. *J Wildl Manag* 46:367–374.

Elton, C. S., and M. Nicholson. 1942. The ten-year cycle in numbers of the lynx in Canada. *J Anim Ecol* 11:215–244.

Graham, S. 1987. The changing role of animal dealers. In *AAZPA Annual Conference Proceedings*, 646–652. Wheeling, WV: American Association of Zoological Parks and Aquariums.

Graham, S. 1996. Issues of surplus animals. In *Wild Mammals in Captivity*, ed. D. G. Kleiman, M. E. Allen, K. V. Thompson, and S. Lumpkin, 290–296. Chicago: University of Chicago Press.

Hanson, E. 2002. *Animal Attractions*. Princeton: Princeton University Press.

Hutchins, M. 2001. Animal welfare: what is AZA doing to enhance the lives of captive animals? In *AZA Annual Conference Proceedings*, 117–129. Silver Spring, MD: American Zoo and Aquarium Association.

Jewell, P.A., and S. Holt. 1981. *Problems in Management of Locally Abundant Wild Mammals*. New York: Academic Press.

Kirkwood, J. K. 2001. Welfare, animal. In *Encyclopedia of the World's Zoos*, Vol. 3, ed. C. E. Bell, 1320–1324. Chicago: Fitzroy Dearborn.

Kisling, V. N. Jr. 2001. History. In *Encyclopedia of the World's Zoos*, Vol. 2, ed. C. E. Bell, 565–570. Chicago: Fitzroy Dearborn.

Lacy, R. 1991. Zoos and the surplus problem: an alternative solution. *Zoo Biol* 10: 293–297.

Lacy, R. 1995. Culling surplus animals for population management. In *Ethics on the Ark*, ed. B. G. Norton, M. Hutchins, E. F. Stevens, and T. L. Maple, 187–194. Washington, DC: Smithsonian Institution Press.

Lindburg, D. G. 1991. Zoos and the "surplus" problem. *Zoo Biol* 10:1–2.

Lindburg, D., and L. Lindburg. 1995. To cull or not to cull. In *Ethics on the Ark*, ed. B. G. Norton, M. Hutchins, E. F. Stevens, and T. L. Maple, 195–208. Washington, DC: Smithsonian Institution Press.

Liu, J., G. C. Daily, P. R. Ehrlich, and G. W. Luck. 2003. Effects of household dynamics on resource consumption and biodiversity. *Nature* (Lond) 421:530–533.

Maple, T., R. McManamon, and E. Stevens. 1995. Animal care, maintenance, and welfare. In *Ethics on the Ark*, ed. B. G. Norton, M. Hutchins, E. F. Stevens, and T. L. Maple, 219–234. Washington, DC: Smithsonian Institution Press.

McAlister, E. 2001. Ethics. In *Encyclopedia of the World's Zoos*, Vol. 1, ed. C. E. Bell, 429–431. Chicago: Fitzroy Dearborn.

Mench, J. A., and M. D. Kreger. 1996. Ethical and welfare issues associated with keeping wild mammals in captivity. In *Wild Mammals in Captivity*, ed. D. G. Kleiman, M. E. Allen, K. V. Thompson, and S. Lumpkin, 5–15. Chicago: University of Chicago Press.

Norton, B. G., M. Hutchins, E. F. Stevens, and T. L. Maple, ed. 1995. *Ethics on the Ark*. Washington, DC: Smithsonian Institution Press.

Piacsek, B. E. 1987. Effects of nutrition on reproductive endocrine function. In *Handbook of Endocrinology*, ed. G. H. Gass and H. M. Kaplan, 143–151. Boca Raton, FL: CRC Press.

Schaffer, N. 2001. Breeding and reproduction. In *Encyclopedia of the World's Zoos*, Vol. 1, ed. C. E. Bell, 178–183. Chicago: Fitzroy Dearborn.

Seal, U. S., R. Barton, L. Mather, K. Oberding, E. D. Plotka, and C. W. Gray. 1976. Hormonal contraception in captive female lions (*Panthera leo*). *J Zoo Anim Med* 7: 12–20.

Symanski, R. 1996. Dances with horses: lessons from the environmental fringe. *Conserv Biol* 10:708–712.

Terman, C. R. 1973. Reproductive inhibition in asymptotic populations of prairie deermice. *J Reprod Fertil Suppl* 19:457–463.

Walker, S. R. 2001. Acquisition, animal. In *Encyclopedia of the World's Zoos*, Vol. 1, ed. C. E. Bell, 1–3. Chicago: Fitzroy Dearborn.

Whittaker, W. J. 2001. Pet trade. In *Encyclopedia of the World's Zoos*, Vol. 1, ed. C. E. Bell, 997–1000. Chicago: Fitzroy Dearborn.

Wiese, R., and M. Hutchins. 1994. *Species Survival Plans: Strategies for Wildlife Conservation*. Bethesda, MD: American Zoo and Aquarium Association.

Wilson, G., N. Dexter, P. O'Brien, and M. Bomford. 1992. *Pest Animals in Australia: A Survey of Introduced Wild Mammals*. Canberra: Bureau of Natural Resources.

Wolfe, M. L. 1986. Population dynamics of feral horses in western North America. *J Equine Vet Sci* 6:231–235.

Wuichet, J., and B. Norton. 1995. Differing conceptions of animal welfare. In *Ethics on the Ark*, ed. B. G. Norton, M. Hutchins, E. F. Stevens, and T. L. Maple, 235–250. Washington, DC: Smithsonian Institution Press.

Xanten, W. A. Jr. 2001. Disposition. In *Encyclopedia of the World's Zoos*, Vol. 1, ed. C. E. Bell, 368–371. Chicago: Fitzroy Dearborn.

Part I

THE ISSUES

The issues involved with the use of contraception in captive and free-ranging wildlife go well beyond the realm of biology. Obviously, understanding the biological aspects of contraception, such as efficacy, safety, reversibility, and whether they differ among taxa, is essential for planning a program that uses contraception as one of the methods of population control. Indeed, the majority of this book is devoted to biological issues. However, Chapters 1 and 2 discuss other variables that play a role in the development of a contraception program, specifically, the ethical concerns that are raised and the legal issues that must be addressed.

The ethical questions regarding the use of contraception to control wildlife populations are related to the value humans place on other living creatures and our responsibility toward them. There are differences of opinion even among those people who are concerned with the welfare of individual animals, based on different definitions of what constitutes animal welfare and well-being. Although people's values may shape their definition of animal well-being, research and monitoring are required to inform people whether their proposed solutions to population control actually conform to their values. The importance of an informed public is clear when one realizes how profoundly public opinion influences wildlife policy and regulations.

The labyrinth of regulations that face scientists and animal managers interested in contraceptive research or application is detailed in Chapter 2. From the development of a new product to the application of that product in the field, ignorance of relevant regulations can bring work with a promising contraceptive method to a complete halt. These chapters present basic ethical considerations and outline the major layers of bureaucracy that impact the structure and function of wildlife contraception programs.

INGRID J. PORTON

1
THE ETHICS OF WILDLIFE CONTRACEPTION

Contraception is a management tool that is utilized to control populations of captive and free-ranging wildlife; however, its use in either type of population is not universally supported. A discussion of the ethics of wildlife contraception is integrally woven into the issue of surplus animals, overpopulation, and our increasing ability to intervene, which in turn centers on the values people place on wildlife and their perceptions of animal welfare (Norton et al. 1995; Kellert 1996). In this chapter, I discuss issues raised by proponents and opponents of wildlife population control as they apply to mammals.

FREE-RANGING WILDLIFE

Free-ranging wildlife populations are regulated naturally by food abundance, predation, parasites, and disease. Removal or a change in the balance of one or more of these limiting factors can lead to overpopulation, which can have negative consequences for many species and their habitats. Overpopulation can result from accidental or purposeful introductions of nonendemic species into habitats where no regulating mechanisms have evolved. Even when they are within the range of historical numbers, populations have also been deemed excessive when they compete for resources desired by people or when they present a risk to human health or safety.

Governmental wildlife management agencies, whether at the federal, state, or local level, are the regulatory bodies responsible for problems of wildlife overpopulation. By definition, wildlife management involves "changing the characteristics and interactions of habitats, wild animal populations, and men to achieve specific human goals by means of the wildlife resource" (Giles 1971). When human goals come into conflict with wildlife, population control measures are enacted. The means by which population control is achieved is strongly influenced by public opinion, and public opinion has changed over the years (Gill and Miller 1997; Curtis et al. 1997). Historically, hunting has been the predominant form of population control, but a shift in public attitude began after World War II when a proanimal, antihunting sentiment corresponded to the rise of the humane, feminist, and peace movements (Kellert 1996). Public opinion now includes a strong voice for animal protection, which has led to an increased interest in wildlife contraception as an alternative to hunting as a means of population control (Cohn 2002).

By the 1970s, increased awareness surrounding the issue of endangered species and the emergence of the environmental movement further contributed to changing attitudes toward wildlife (Kellert 1996). Interested in this phenomenon, Kellert (1996) identified nine basic values that humans ascribe to nature: utilitarian, naturalistic, ecologistic-scientific, esthetic, symbolic, humanistic, moralistic, dominionistic, and negativistic. Learning, experience, and the culture in which people live all contribute to the development of their values. Thus, the value a person places on wildlife shapes their perception of wildlife management, including contraception.

LETHAL VERSUS NONLETHAL POPULATION CONTROL

Wildlife control methods can be classified into two broad categories: lethal and nonlethal. Lethal methods include hunting, trapping; and poisoning; nonlethal methods include translocation and contraception. Historically, lethal methods have predominated, but the emergence of new contraceptive techniques, especially immunocontraception, has shown that there may be viable and realistic nonlethal alternatives to control population growth in wildlife (Tuyttens and MacDonald 1998; see Chapter 13, Contraception in Free-Ranging Wildlife). The option of contraception has led to renewed discussions about which population control methods are humane and ethical.

There are three value-based positions regarding wildlife population control that could be described as dogmatic. Proponents of the first position reject the precept that nonhuman animals are property. Within this philosophy, the term wildlife management is itself flawed, because it is felt that humans should not in any way manage wildlife populations whether through hunting or contraception.

Both methods are seen as equally abhorrent in that they interfere with an animal's right to live and control its own life. Within this philosophical position, manipulating an animal's reproductive life is as immoral as is taking its life.

The second position holds that hunting is the only correct and natural form of population control. Proponents of this position maintain that nature produces a surplus. Hunting replicates a fundamental biological process designed to control this surplus and, as such, is a part of the life–death cycle. The central moral issue relates to what is best for the environment, that the whole is more important than the individuals that encompass it. Furthermore, this group contends that hunting connects humans to and stimulates an ethical relationship with nature (Kellert 1996; Conservation Force 2001). Some proponents of this position have suggested that contraception and sterilization are not only unnatural but can lead to the extinction of species (Conservation Force 2001).

The third view is that hunting causes pain and suffering to individual sentient beings and should be abolished, especially when nonlethal methods such as contraception are available. Advocates have suggested that opposition to contraception as a method of population regulation exposes the hunters' true self-interests and refutes the idea that their primary concern is the health of the ecosystem. This view holds that if habitat loss or overpopulation necessitates population control to prevent animals from needless suffering, then nonlethal methods are the only ethical option.

Alternatively, many people have mixed views regarding the most appropriate methods of population regulation. For example, a survey of Coloradans showed that the majority of respondents were not opposed to hunting as a wildlife management technique if it benefited the animals, for example, returned the population to the habitat's carrying capacity. However, when asked about the act of hunting, respondents' opinions were predicated on the reason for hunting, with approval highest when the purpose was for obtaining meat, lower when simply for recreation, and lower yet when the aim was to obtain a trophy. The Colorado survey found that a quarter of the people felt that animals had the right to be protected from human exploitation (Gill and Miller 1997).

People do, however, distinguish among species. For example, there are those who oppose hunting but do not object to killing creatures they believe lack sentient abilities (Kellert 1996). Clearly, public opinion is diverse with respect to how wildlife populations should be managed. Because public pressure influences wildlife management policies, it is important for wildlife biologists to provide accurate information on management methods and outcomes that can allow people to judge which options are most consistent with their value systems.

FULFILLING MORAL OBLIGATIONS

For those who are totally opposed to managing wildlife populations, the debate between population control through lethal or contraceptive methods is moot. However, people who believe that the adverse consequences of taking no action are more inhumane than enacting population control are faced with judging which methods are most humane. The reality of which population control method achieves the desired outcome of reduced pain and distress is more complex and contextual than the polemic paradigm of lethal versus nonlethal control. An objective evaluation is dependent on understanding each technique within the two broad categories of lethal and nonlethal. For example, if the objection to lethal methods is that they cause too much pain, are there sufficient differences in the level of animal suffering among the lethal options? Might hunting by an expert sportsman be more humane than trapping or poisoning? Could reintroduction of extirpated predators be more morally acceptable than hunting by humans?

Is pain or distress to the animal associated with the administration or use of contraceptives? Surgical sterilization, although expensive and labor intensive, may be an option in some situations (Tuyttens and Macdonald 1998) but carries a risk of death from anesthesia or complications following infection. Successive infertile reproductive cycles or a complete cessation of cycles following sterilization may cause behavioral changes that could be beneficial, detrimental, or neutral to the animals within the population. For example, male and female white-tailed deer (*Odocoileus virginianus*) alter their activity budget when females experience polyestrous nonconceptive cycles (Guynn 1997). Whether such changes negatively affect deer feeding patterns, energy expenditure, and consequent survival through the winter has not been determined. In contrast, Kirkpatrick and Frank (see Chapter 13, Contraception in Free-Ranging Wildlife) have found that mares prevented from reproducing by porcine zona pellucida (PZP) vaccine are in better condition and live longer than their reproductive counterparts.

Decisions concerning the behavioral implications of population control should also be made within the context of species social organization. If one goal is to select a population control method that causes the least disruption of social relationships, then the social consequences of using contraception versus culling members of social groups must be thoroughly examined. For example, elephants live in complex, matrilineal family groups in which allomothering (caregiving by others in the group) is common, social bonds are strong, and family relationships are often maintained through generations (Moss 1988). In contrast, white-tailed deer live in loose social groups that females leave to give birth, and allomothering does not occur. Are the social ramifications of killing a female elephant different from those of killing a female deer? Would the administration of a sterilizing agent be more disruptive to the stability of a highly social wolf pack in the long term than the culling of individual pack members?

Those concerned about selecting a population control method that causes the least pain and distress also must identify which individual animals are the objects of their concern; this may include evaluating the chance of success among the different methods, because no plan is completely flawless. Even proven techniques may not succeed for reasons of logistical problems unique to the specific situation. One consequence of a failed or only partially successful plan could be that, in the end, a greater number of animals are subjected to unnecessary suffering. Another perspective is that diverting the optimal use of limited resources, such as funds or expert staff, from another program that could benefit more individual animals or species is in itself an ethical choice. All things being equal, if hunting and fertility control are similarly effective at population management, but fertility control drains financial resources that would otherwise be spent on different conservation activities that improve the survival of other animals, which is ultimately of greater value? Which group of individuals has priority? Although the ecological approach, which prioritizes the health of an ecosystem, is often portrayed as being in conflict with the approach that emphasizes individual animal well-being, welfare advocates may be advised to assess the number of individual animals that ultimately benefit from the long-term effects of prioritizing habitat management. Wildlife and animal welfare advocates could also examine how their own lifestyles contribute to habitat loss, for example, through overutilization of natural resources and urban sprawl.

CAPTIVE WILDLIFE

Nondomestic mammals are held in different captive environments, including zoos accredited by regional zoo and aquarium associations, such as the American Zoo and Aquarium Association (AZA, North America), the European Aquarium and Zoo Association (EAZA), and the Australasian Regional Association of Zoological Parks and Aquaria (ARAZPA), and by unaccredited zoos, roadside animal attractions, sanctuaries, research laboratories, animal dealers, various entertainment enterprises such as circuses and animal actor suppliers, and exotic pet owners. Our society also includes a certain proportion of people who completely oppose keeping wild animals in captivity. Although this is a valid view and a subject worthy of ethical discourse (Wagner 1995; Loftin 1995), this chapter does not address that issue because it is not directly related to population management. Instead, the discussion focuses on wildlife in zoo-based breeding programs and the role of contraception in the management of these captive populations. For purposes of this discussion, the term zoo refers only to those that have been accredited by their regional zoo association and does not include unaccredited zoos and roadside animal attractions.

Zoos define their mission as institutions that advance wildlife conservation through the maintenance of captive populations for public education, scientific research, reintroduction, and conservation fund-raising (McAlister 2001; Hutchins 2001). Cooperative breeding programs for many species have been instituted to assure sustainable populations well into the future with minimal need to remove additional animals from the wild. These managed programs strive to maintain genetically, demographically, and behaviorally healthy populations of captive wildlife. Although these populations can serve as genetic reservoirs, should the need ever arise to reintroduce them to the wild, Hutchins (2003) suggests that the role of captive populations for reintroduction will decrease as zoo-sponsored in situ conservation efforts increase. In truth, the role of zoos as environmental education centers that have the potential to stimulate conservation action in visitors is viewed by many zoo professionals as equal to or even more important than the role of zoos as genetic reservoirs.

Cooperatively managed breeding programs (for example, AZA's Species Survival Plan, SSP) develop population master plans that make both breeding and nonbreeding recommendations. This planning is necessary because unlimited breeding of any species in captivity will result in more animals than can be humanely accommodated. Because space in zoos is limited (Ballou 1987; Xanten 2001) and because zoos are interested in maximizing their conservation and education goals, zoo population biologists have developed models to calculate the number of individuals required to maintain a self-sustaining population (Ballou and Foose 1996). The calculations account for the number of wild-born animals that founded the captive population (that is, the genetic diversity of the population) and the subsequent breeding history of the population. As a result, the population size required to achieve the stated goals (target population size) may vary even among species with similar life history characteristics.

Zoos must regulate captive population growth to meet but not exceed the calculated target population size. It is here that opinions vary concerning the best approach to control population growth. Regulation can include limiting breeding through the use of reversible contraception, permanent sterilization, the euthanasia of surplus animals, or the transfer of individuals no longer needed for breeding programs (Asa et al. 1996; Graham 1996; Xanten 2001). In reality, most zoo biologists do not believe that the surplus animal problem can be solved with only one approach. The ethical quandaries center on which options are most appropriate under what circumstances.

CONTRACEPTION

Reversible contraception has become an increasingly available option for fertility control in exotic mammals since the mid-1970s when research on the efficacy of the melengestrol acetate (MGA) implant in exotic felids was initiated (Seal et al. 1976; Porton et al. 1990). The need for contraceptive options in addition to the obvious choice of separating males from females was, and continues to be, viewed as important by many zoo animal managers because it addresses the social and behavioral needs of the animals as well as the problem of limited housing. Reversible contraception is seen as particularly beneficial in social species because it eliminates the socially disruptive procedure of removing individuals (male or female) during estrus without sacrificing future fertility. It also allows better utilization of housing, for example, by permitting offspring to remain in natal groups without the risk of inbreeding and by not having to find additional space to house individuals not recommended to breed. Advocates see contraception as the most responsible solution for the prevention of overpopulation within captive populations (Asa et al. 1996; Lacy 1995).

While recognizing the value of reversible contraception, some individuals express concern over the possibility that future research may discover new health risks correlated to contraceptive use. Although more data are available on the safety or behavioral consequences of contraception in some species than in others, it remains true that all reversible contraceptive methods used in zoo mammals are experimental and that long-term research on safety, toxicity, and carcinogenesis is required for all species. Without conclusive data on the consequences of contraceptive use in all species, the ethical dilemma becomes whether the possibility of health risks associated with contraceptive use is outweighed by the benefit of not separating animals or of not producing surplus animals. Which risk is more acceptable, the possibility of a drug-induced side effect in the contracepted parent that could shorten its life, or the possibility that the well-being of the offspring may be compromised by overcrowding or permanent transfer to a substandard facility? Based on the increased use of contraception in a variety of mammalian species (Contraception Database), it appears that many managers have decided that taking some reasonable risks is more principled than producing surplus animals.

Some zoo professionals object to the use of contraceptives based on the belief that preventing animals from mating and raising young deprives those animals of a fundamental and enriching part of their life. These zoo professionals argue that it is wrong, and even hypocritical, to emphasize the importance of developing enriched captive environments that facilitate performance of natural behaviors while at the same time advocating the prevention of natural reproductive behavior. This view holds that all social aspects of mating and rearing offspring are of overriding importance to the well-being of captive animals and to prevent this experience could be considered unethical (Glatston 1998; Holst 1998; Wiesner and Maltzan 2000). Many who espouse this position do agree that zoos should only carry out a limited number of breeding programs but believe that within that limit the individuals in the programs should be allowed to breed. Rather than using contraception to prevent overpopulation, proponents advocate that surplus offspring should be euthanized. To provide the maximum social benefit to the parents and other group members, the offspring should be removed and euthanized at the age at which they would normally disperse from their group (Holst 1998).

Another position expressed by some who oppose contraception or advocate its limited use is that zoos serve to educate the public about the natural lives of animals. Strict limits on reproduction will deprive many zoo visitors of the educational opportunities to observe not only parenting behaviors but also the maturation process of the offspring. Persons advocating this position may be unaware of or have not thought through the consequences of unregulated reproduction.

EUTHANASIA

The topic of euthanasia is directly relevant to a discourse on the ethics of contraceptive use, because euthanasia is one solution for the removal of surplus animals. Euthanasia has always been a recommended option for those animals that are seriously or terminally ill (AVMA 1993). A more controversial view, yet one supported by some zoo professionals, is that individuals housed in less than adequate facilities, especially singly housed individuals of a social species with dim prospects for future companionship, should be humanely euthanized.

As already mentioned, some advocate euthanasia over contraception as the solution that addresses both animal welfare concerns and overpopulation. This management philosophy maintains that the behavioral health of the reproductive individual is more important than the lives of the offspring. This opinion was expressed by Holst (1998) when he said: "By not accepting euthanasia of surplus animals, we accept a distressing deterioration in animal welfare instead, namely for those animals that are not allowed to breed. We should not forget that we work with individual animals that should be given a reasonable life. They are not just numbers. And having young is a social enrichment that really matters." The author goes on to say that death by euthanasia is less painful than the deaths animals typically endure in the wild. Opponents of this approach argue that a "comprehensive welfare concept includes the opportunity to live for a natural period of time" (Margodt 2000).

Another point of view on the role of euthanasia in zoo animal management was expressed by Lacy (1991, 1995), who advocates contraception as the most responsible manner by which to avoid the production of excessive surplus. Lacy contends that even the best managed breeding program contains individuals that can no longer be absorbed into available housing without taking space away from individuals that can and should contribute positively to the program's objectives. If the objectives are to maintain a self-sustaining captive population as insurance against species extinction, or for other conservation objectives, then the health of the population supersedes extending the life of a genetically surplus individual. Thus, in this argument, contraception is advocated as the most acceptable method to ensure that target population goals are met. In fact, Lacy (1995) forcefully states that "the unnecessary production of surplus animals is an inexcusable dereliction of our duty to use resources wisely for the protection of the natural world." Population regulation that cannot be accomplished solely through contraception should then be done through euthanasia.

ANIMAL TRANSFERS

Animals within a coordinated breeding program, such as an SSP, are regularly transferred among participating zoos. Such transfers are essential to the operation of a genetically managed breeding program. The problems and concerns related to animal transfers center around individuals that are moved to facilities outside the regional zoo association's cooperative program (Brouwer 1993; Lindburg 1991). For example, those zoo associations that have developed animal caretaking standards discourage member institutions from transferring animals to facilities that do not meet these standards. AZA has gone so far as to include the transfer of animals to inadequate facilities as a violation of the association's Professional Code of Ethics (AZA 2000). Critics say that these standards, although appearing earnest and responsible, are too vague and continue to allow many zoo-bred animals to ultimately arrive at facilities that do not meet their welfare needs (Goldston 1999; Green 1999).

There are records of transferred zoo-bred mammals from big cats, lemurs, and chimpanzees to a range of ungulate species that have eventually reached roadside animal attractions, circuses, for-profit animal breeding facilities, exotic animal auctions, and the pet trade. Even when such animals are not directly transferred to substandard facilities, individual animals or their progeny may subsequently be transferred to inadequate situations (Xanten 2001). The extent of the problem is quite significant (Goldston 1999; Green 1999). For example, there are reputed to be more tigers in private ownership than there are in the wild (Peterson 2002). An unknown but not insignificant number of unwanted exotic animals ultimately end in unacceptable conditions that are considered inhumane.

The core of this issue is not so much whether contraception should be used in a breeding program but to what extent it should be used. Some propose that more animals should be allowed to breed if there are facilities outside accredited zoos that are interested in obtaining surplus animals. The discussion encompasses who has the right to own exotic mammals, who defines welfare standards, and to what degree zoos are responsible for the fate of the animals that leave their care and ownership. Some zoo professionals argue that the responsibility ends when the animal is placed in a facility that meets their animal care standards and that the zoo is not accountable for subsequent transfers. Others proclaim that zoos are ethically responsible for all subsequent transfers of the original animal as well as all its progeny. They further reason that, because there are known cases in which surplused zoo animals or their offspring have eventually reached facilities with deplorable conditions, the only unequivocal solution is to accept cradle-to-grave responsibility for all zoo-bred animals. Although more scientific studies are required to substantiate what environmental and social housing standards are sufficient to meet the well-being of the exotic species housed in zoos, it is indisputable that once outside the control of accredited institutions, a certain percentage of individuals born in zoos or their offspring will endure what can only be termed substandard conditions. Accredited zoos bring institutional support to the veterinary, nutritional, and behavioral care of animals. Is it not reasonable to assume that sending reproductively intact animals out of a cooperative breeding program ultimately can result in a greater number of individuals suffering than would have resulted had the original zoo euthanized that animal or maintained it in less than ideal housing? Is it morally justifiable to give up responsibility for monitoring and managing an animal's life and death on the chance that the animal could be well cared for at an unknown facility?

ZOO VALUES AND MISSION

The ethics of contracepting captive mammals is ultimately tied to a person's value system. That value system is also the foundation from which the mission of a captive breeding program is evaluated. Within the zoo community, a pivotal document that articulates the shared values that institutions place on the existence of cooperative wildlife management programs is the Regional Collection Plan (RCP).

The RCP details the mission for maintaining captive populations of selected species, then outlines the goals and objectives to achieve this stated mission. It is a difficult document to produce because it requires a level of collective introspection to judge whether the recommended programs adequately reflect the moral justification for maintaining the captive populations. It is a large responsibility because the resulting decisions concerning the form and function of the programs have consequences both for species conservation and for individual animal welfare.

The paramount mission articulated in many RCPs is the conservation of species in their natural habitat. Experts point out that there are multiple ways in which captive populations can contribute to wildlife conservation, but there are insufficient data to judge which programs have or will have the greatest impact. Is the contribution of preserving a species in captivity for possible reintroduction greater than maintaining a charismatic species that may so capture public interest as to possibly influence the direction of public policy? As such questions are yet unanswered, one approach to selecting captive animal programs may be to identify which programs can achieve the greatest or equal conservation benefits while simultaneously causing the fewest compromises to individual well-being.

Even if the foregoing selection process maximizing individual and species welfare is adopted, zoo professionals are still faced with the challenge of defining individual animal well-being. Is individual animal well-being better achieved by preventing the production of surplus animals through contraception? Is it more ethical to forego birth control and thereby provide an enriched life for a subset of captive animals and euthanize all others? Is it more ethical to euthanize surplus animals than to send them to unaccredited facilities where the long-term welfare of the individual or its offspring cannot be guaranteed?

The AZA Contraception Advisory Group was initially formed because of the manifest reality that zoos exist and animals reproduce. Humans have brought animals into captivity and in doing so, we believe, must assume the responsibility of caring for them. We concur with Norton's (1995) statement: "Once they (wild animals) are made part of the mixed communities of zoos and captive breeding programs, we accept expanded responsibility for their welfare, and our obligation will be governed by the comparative level of self-awareness and complexity that constitutes their ontological good." We believe contraception is an indispensable tool for preventing the creation of more animals than can be housed in facilities that share our caretaking standards while, at the same time, maintaining the reproductive capability of individuals that comprise conservation breeding programs. We believe that the use of reversible contraception is ethically justified within the moral value system of many people, and that developing safe and effective contraceptive choices not only contributes to the ethical management of wildlife but is a moral obligation.

FINAL REMARKS

The ethical use of wildlife contraception requires that it be applied wisely with the aim of providing the greatest benefit for the welfare and survival of both individual animals and species. To accomplish this goal, immediate and long-term consequences of all methods used to control wild or captive populations must be weighed. The unknown should not be used as an excuse to abdicate responsibility for possible outcomes. The hoped-for outcome will be based on a person's value system, and there will always be fundamental differences in the value people place on wildlife. It is, however, incumbent on people to ensure that a proposed solution to wildlife overpopulation does, in actuality, reflect their values.

ACKNOWLEDGMENTS

The author thanks Joe Bielitzki, Scott Carter, Lisa Kelley, and Alice Seyfried for their thoughtful review of the manuscript.

REFERENCES

Asa, C. S., I. Porton, A. M. Baker, and E. D. Plotka. 1996. Contraception as a management tool for controlling surplus animals. In *Wild Mammals in Captivity*, ed. D. G. Kleiman, M. E. Allen, K. V. Thompson, and S. Lumpkin, 451–467. Chicago: University of Chicago Press.

AVMA. 1993. Report of the American Veterinary Medical Association panel on euthanasia. *J Am Vet Med Assoc* 202 (2): 229–249.

AZA. 2000. *AZA Accession/De-accession Policy*. http://www/aza.org/About AZA/AD Policy. Silver Spring, MD: American Zoo and Aquarium Association.

Ballou, J. 1987. Small populations, genetic diversity and captive carrying capacities. In *Annual Proceedings of the AAZPA*, 33–47, September 20–24, 1987, Portland, OR.

Ballou, J. D., and T. J. Foose. 1996. Demographic and genetic management of captive populations. In *Wild Mammals in Captivity*, ed. D. G. Kleiman, M. E. Allen, K. V. Thompson, and S. Lumpkin, 263–283. Chicago: University of Chicago Press.

Brouwer, K. 1993. EAZA/EEP available and wanted list; a step further towards responsible animal exchanges in European zoos. In *EEP Yearbook 1992/93*, ed. L. E. M. DeBoer, K. Brouwer, and S. Smits, 278–284. Amsterdam: EAZA/EEP Executive Office.

Cohn, P. N. 2002. Wildlife contraception: an introduction. www.animal-law/hunting/contrintro.htm.

Conservation Force. 2001. Big changes announced at CIC meeting. www.conservation-force.org/info/get_hunter_art.cfm?art_id=52.

Curtis, D., D. J. Decker, R. J. Stout, M. E. Richmond, and C. A. Loker. 1997. Human dimensions of contraception in wildlife management. In *Contraception in Wildlife Management*, ed. T. J. Kreeger, 247–255. USDA APHIS Technical Bulletin 1853. Washington, DC: US Department of Agriculture.

Giles, R. H. 1971. The approach. In *Wildlife Management Techniques*, ed. R. H. Giles, 3rd ed., 1–4. Washington, DC: The Wildlife Society.

Gill, R. B., and M. W. Miller. 1997. Thunder in the distance: the emerging policy debate over wildlife contraception. In *Contraception in Wildlife Management*, ed. T. J. Kreeger, 257–267. USDA APHIS Technical Bulletin 1853. Washington, DC: US Department of Agriculture.

Glatston, A. R. 1998. The control of zoo populations with special reference to primates. *Anim Welf* 7:269–281.

Goldston, L. 1999. The animal business (four-part series of articles). San Jose, CA: *San Jose Mercury News*, February 7, 8, 9, 10.

Graham, S. 1996. Issues of surplus animals. In *Wild Mammals in Captivity*, ed. D. G. Kleiman, M. E. Allen, K. V. Thompson, and S. Lumpkin, 290–296. Chicago: University of Chicago Press.

Green, A. 1999. *Animal Underworld*. New York: Public Affairs.

Guynn, D. C. Jr. 1997. Contraception in wildlife management: reality or illusion. In *Contraception in Wildlife Management*, ed T. J. Kreeger, 241–245. USDA APHIS Technical Bulletin 1853. Washington, DC: US Department of Agriculture.

Holst, B. 1998. Ethical costs in feeding and breeding procedures. In *EEP Yearbook 1996/97*, ed. F. Rietkerk, S. Smits, and M. Damen, 453–454. Amsterdam: EAZA/EEP Executive Office.

Hutchins, M. 2001. Animal welfare: what is AZA doing to enhance the lives of captive animals? In *AZA Annual Conference Proceedings*, 117–129, September 7–11, 2001, Saint Louis, MO.

Hutchins, M. 2003. Zoo and aquarium animal management and conservation: current trends and future challenges. *Int Zoo Yearb* 38:14–28.

Kellert, S. R. 1996. *The Value of Life*. Washington, DC: Island Press.

Lacy, R. 1991. Zoos and the surplus problem: an alternative solution. *Zoo Biol* 10: 293–297.

Lacy, R. 1995. Culling surplus animals for population management. In *Ethics on the Ark*, ed. B. G. Norton, M. Hutchins, E. F. Stevens, and T. L. Maple, 187–194. Washington, DC: Smithsonian Institution Press.

Lindburg, D. G. 1991. Zoos and the "surplus" problem. *Zoo Biol* 10:1–2.

Loftin, R. 1995. Captive breeding of endangered species. In *Ethics on the Ark*, ed. B. G. Norton, M. Hutchins, E. F. Stevens, and T. L. Maple, 164–180. Washington, DC: Smithsonian Institution Press.

Margodt, K. 2000. *The Welfare Ark*. Brussels: Vrije Universiteit Brussel (VUB) University Press.

McAlister, E. 2001. Ethics. In *Encyclopedia of the World's Zoos*, Vol. 1, ed. C. E. Bell, 429–431. Chicago: Fitzroy Dearborn.

Moss, C. J. 1988. *Elephant Memories*. New York: Morrow.

Norton, B. G. 1995. Caring for nature. In *Ethics on the Ark*, ed. B. G. Norton, M. Hutchins, E. F. Stevens, and T. L. Maple, 102–121. Washington, DC: Smithsonian Institution Press.

Norton, B. G., M. Hutchins, E. F. Stevens, and T. L. Maple, ed. 1995. *Ethics on the Ark*. Washington, DC: Smithsonian Institution Press.

Peterson, I. 2002. Cute pet to some, untamed killer to others. Tigers are increasingly roaming through backyards and back streets. *New York Times* Feb. 1, 2002.

Porton, I., C. Asa, and A. Baker. 1990. Survey results on the use of birth control methods in primates and carnivores in North American zoos. In *AAZPA Annual Conference Proceedings*, 489–497, September 23–27, 1990, Indianapolis, IN.

Seal, U. S., R. Barton, L. Mather, K. Oberding, E. D. Plotka, and C. W. Gray. 1976. Hormonal contraception in captive lions (*Panthera leo*). *J Zoo Anim Med* 7:1–17.

Tuyttens, F. A. M., and D. W. MacDonald. 1998. Fertility control: an option for nonlethal control of wild carnivores? *Anim Welf* 7:339–364.

Wagner, F. 1995. The should or should not of captive breeding: whose ethic? In *Ethics on the Ark*, ed. B. G. Norton, M. Hutchins, E. F. Stevens, and T. L. Maple, 209–214. Washington, DC: Smithsonian Institution Press.

Wiesner, H., and J. Maltzan. 2000. Population control in Bavarian zoos. In *Proceedings, European Association of Zoo and Wildlife Veterinarians*, 77–81, May 31–June 4, 2000, Paris, France.

Xanten, W. A. Jr. 2001. Disposition. In *Encyclopedia of the World's Zoos*, Vol. 1, ed. C. E. Bell, 368–371. Chicago: Fitzroy Dearborn.

JAY F. KIRKPATRICK AND WOLFGANG JÖCHLE

2
REGULATORY ISSUES

REGULATORY ISSUES RELATED TO CONTRACEPTIVE DRUG PRODUCTION

Contraceptive products formulated for wildlife, either chemical compounds formulated as pharmaceutical preparations (that is, drugs) or vaccines, require prescriptions by a licensed veterinarian. Worldwide, prescription items are controlled by regulatory agencies at a national level (for example, United States) or supranational level (for example, European Union, EU). These agencies approve the sale of these drugs and vaccines, based on demonstrations of efficacy, safety, and the capability to produce them consistently under good manufacturing practices. In some countries, this approval for sale is called registration. The requirements for approval for sale or registration are not yet globally uniform, but ongoing attempts of harmonization have yielded some progress regarding manufacturing practices. For our readers, regulatory conditions prevailing in North America (United States and Canada) and in the EU are of primary interest.

In the United States, animal drugs and those vaccines influencing physiological processes usually are under the regulatory jurisdiction of the Center for Veterinary Medicine (CVM) of the Food and Drug Administration (FDA), an agency of the Department of Health and Human Services (DHHS) of the federal government. In the United States, vaccines indicated for preventing infectious diseases (biologicals) are regulated by the Animal and Plant Health Inspection Service (APHIS) of the US Department of Agriculture (USDA). Erroneously, the assumption is often made that any vaccine designed to prevent fertility automatically would be handled by APHIS, thus avoiding FDA.

The CVM/FDA and APHIS/USDA have signed a Memorandum of Understanding regarding vaccines, which includes contraceptive vaccines. Sponsors are asked to file applications to both agencies, explaining the product and label claims. The agencies will subsequently interact and decide which will take the regulatory responsibility for the biological product's approval. After the agencies decide which of them will assume responsibility for oversight of the product, the sponsor should request a conference with that agency for review of the intended developmental studies and to agree on the data needed on efficacy, safety, and manufacturing process. These data requirements may differ somewhat between CVM/FDA and APHIS/USDA. An example of such a contraceptive product is an antigonadotropin-releasing hormone (anti-GnRH) vaccine that was recently approved for sale by APHIS/USDA for use in male dogs.

Sponsors for the development of any new contraceptive animal health drug or vaccine intended for approval by CVM/FDA for sale must, as a first step, file an Investigational New Animal Drug Application (INADA) with the CVM (www.fda.gov/cvm/guidance/published.htm). An INADA allows for the interstate shipment of the new drug, which is essential for conducting the required efficacy and safety studies. If the new drug is used in animals that may produce or may become human food (which includes deer, elk, bison, and waterfowl), the product also may be subject to time-consuming and expensive safety and residue studies. All drugs may be evaluated for their potential environmental impact, although this latter requirement can often be waived by requesting an exemption from performing an assessment of environmental impact. Similar steps must be taken with APHIS/USDA.

If this INADA phase is successfully concluded (usually after 3 to 6 years), that is, the drug is efficacious and safe and can be produced to FDA standards, the sponsor can file a NADA (New Animal Drug Application). CVM/FDA has been entitled to charge a user fee for granting a new product approval. A waiver has been granted for products intended only for minor species, which includes wildlife. However, the road to approval still is filled with hurdles that could take several more years.

The regulation of animal health drugs in Canada is the responsibility of the Veterinary Drugs Directorate (VDD), a division of Health Canada's Health Products and Food Branch. The approval process in Canada closely resembles that of the United States. Data on safety, efficacy, and manufacturing generated in the United States are usually accepted, although the VDD often requires that at least one pivotal clinical study be conducted in Canada. In contrast to the situation in the United States, Canada has a pay-as-you-go system in which every step in the approval process generates fees, which are due at the time of submission.

The EU drug approval system is different: sponsors can elect to file for approval either for the entire EU with the Committee for Veterinary Medicinal Products (CVMP) in London, or with 1 of the 18 national regulatory agencies. However, for new chemical ingredients that have never before been used, the compound must pass a safety assessment of the central EU agency (CVMP). Based on data and information the sponsor has to provide and the species in which the drug will be used, the compound will be assigned to lists (Schedules) I to IV, which determines the extent of animal, handler, and consumer safety data required for approval. As outlined earlier, the sponsor can clear further hurdles toward approval from here either nationally (through the regulatory agency in each individual country) or EU wide (through the CVMP). In either approach, the agency's approval will be circulated to all EU member states, which can deny approval individually within 90 days. The kind of data required in the EU is similar to that needed in the United States, but the information has to be presented in a different fashion, and the data are evaluated primarily by experts and committees appointed by the agencies. The experts' recommendations to accept or reject a new drug are usually accepted, but evaluation of a new drug file is costly, irrespective of approval or rejection.

A sponsor's decision to take these roads to the approval of a new drug is based on feasibility: does the anticipated return justify the investment in time, money, and manpower? In the case of contraceptive drugs and vaccines for wildlife or zoo animals, the answer is clearly negative. The United States provides a somewhat abbreviated route under the general umbrella of "minor species" (Center for Veterinary Medicine 1999). However, in the EU the law does not allow "orphan drug status," that is, the recognition that for economical reasons an abbreviated, inexpensive approval process is needed.

A different avenue to new or existing contraceptive drugs or vaccines, approved for use in humans or domestic animals, has been recently opened in the United States via a procedure called extra-label drug use. This path allows veterinarians to prescribe such drugs for other species, provided that such usage is not specifically prohibited (Center for Veterinary Medicine 1994), is safe, and does not create tissue residues in animal products destined for human consumption. Another avenue to help ensure availability of old or new drugs for wildlife and zoo animals is the emergence of compounding pharmacies, which, based on prescriptions, may provide such drugs in dosage forms more suitable for the treatment of nondomestic animals.

Recently, CVM made it known that research involving unapproved drugs in non-food-producing animals may proceed without notification of the agency, if this drug will not be the subject of a future NADA. Precautions regarding safety and the avoidance of food residues are required, and detailed records of drug shipments and drug effects are desirable.

Sponsors interested in the use of an approved animal health drug in zoo animals or wildlife species for which it has not been specifically approved can obtain an INADA from CVM for the intended drug use. Presently, melengestrol acetate (MGA), both in implant form and as a feed additive for zoo animals, is available under this umbrella. Individual investigators may obtain an INADA for the same purpose from CVM.

In Canada and in the EU, national regulatory agencies can be petitioned to import drugs available elsewhere but not yet in their country. Conditions of use and precautions taken must be outlined in detail, and a user fee might be required.

CONTRACEPTION AND ANIMAL WELFARE

In the United States, animals being used in research are covered under the Animal Welfare Act, which requires that research be conducted under protocols approved by an Institutional Animal Care and Use Committee (IACUC). Guidelines can be found in publications such as those of the National Research Council (1996), the National Institutes of Health for the Public Health Service (1996), and the Federation of Animal Science Societies (1999). Testing of new contraceptive methods in many cases requires IACUC review and approval.

In Europe, a number of countries have recently passed revisions of animal welfare laws, which explicitly ban surgical procedures in all animals unless they are medically indicated. Technically, this includes contraceptive procedures such as ovariectomy, ovariohysterectomy, castration, vasectomy, and penile deviation. Governments had to be petitioned to allow spaying and neutering in pets with the argument that these procedures have prophylactic medical value. In carnivores, ovariectomy can forestall the occurrence of uterine pathology and mammary tumors in female carnivores. Castration can lessen the chances of prostate hypertrophy and tumors and ameliorate aggression in male dogs. In male cats, castration can also prevent house soiling (spraying of urine), which is sometimes a cause for euthanasia.

Castration in pigs is under attack in Europe; the traditional cutting of week-old male piglets might be banned soon. Male calves are no longer castrated in Europe. Surgical removal of gonads in any animal might become outlawed eventually. Hence, chemical (pharmaceutical) and biological, that is, vaccine-based, contraception, to which this book is dedicated, may come to play a more important role in management of domestic animals as well as wildlife.

CAPTIVE WILDLIFE

The issue of local regulations for contraceptive use in zoos is less complicated than for free-ranging wildlife, for many obvious reasons. The ownership of the animals rests with the zoos and not with state or federal entities, except the few that are in US Fish and Wildlife Service or National Park Service endangered species recovery programs (for example, black-footed ferret, *Mustela nigripes*; red wolf, *Canis rufus*; Mexican wolf, *Canis lupus baileyi*). Certainly none of the animals in zoos can be considered food animals, and all drugs administered to collection animals are under the direct supervision and on the order of the licensed zoo veterinarian.

Under the Federal Animal Welfare Act zoos are included under the provision of Class "C" (exhibitor). However, the precise regulations for the use of experimental drugs or nonlabel use of drugs for most captive exotic species is not specifically addressed in this act. The Act states: "*Animal* means any live or dead dog, cat, nonhuman primate, guinea pig, hamster, rabbit, or any other warm blooded animal, which is being used for research, teaching, testing, experimentation, or exhibition purposes, or as a pet." Thus, the use of experimental contraceptive drugs in warm-blooded zoo animals nominally falls under the Animal Welfare Act, and that would include the use of noncommercial drugs being used experimentally. Accordingly, this interpretation requires that zoos provide oversight through an Institutional Animal Care and Use Committee (IACUC) or a similar body before experimental contraceptive drugs are applied to collection animals. Any contraceptive drug or vaccine that is registered under an FDA-issued Investigational New Animal Drug Application (INADA) or a New Animal Drug Application (NADA) falls within this interpretation. Most zoos already comply, although the oversight committees are often given names other than IACUC, such as animal care committee. This institutional responsibility is in addition to compliance with other federal regulations regarding the use of experimental drugs, through use under an INADA or NADA as discussed earlier.

FREE-RANGING WILDLIFE

One of the most poorly understood aspects of free-ranging wildlife contraception is the regulatory process. The regulatory oversight for the testing and application of contraceptive drugs becomes considerably more complicated with free-ranging wildlife because of issues such as ownership of animals, legal authority for managing the animals, classification of some species (such as the white-tailed deer,

Odocoileus virginianus) as food animals, and the often confusing management authority between state and federal entities. Depending upon the species in question and the location of that species, one or more regulatory agencies at the federal, state, or local level may be involved. Regardless of species or location, any experimental contraceptive drug for use in free-ranging wildlife that crosses state lines and jurisdictions must adhere to the FDA-CVM regulations regarding experimental drugs, that is, INADA or NADA requirements. Research in general with free-ranging wildlife is exempt from the federal Animal Welfare Act and therefore from oversight by the institutional IACUC. Within academic institutions engaging in animal research, there must be an IACUC, and in most but not all of these universities, the institutions have chosen to require IACUC approval for research with free-ranging animals regardless of whether they legally fall within the domain of the Animal Welfare Act. This decision is simply a responsible action on the part of the institutions, which seek to meet the intent of the Animal Welfare Act rather than the letter of the law.

The "ownership" of the species is of vital importance, and therefore the nature of the environment will dictate the regulatory issues. Historically, nonmigratory wildlife not living on federal land is legally under the management authority of state fish and wildlife departments on both nonfederal public and private land. Thus, a potential wildlife contraceptive project on nonfederal land must have approval by the particular state wildlife agency before any additional steps can take place. The two requirements most often quoted by state wildlife agencies are prior "approval" by the FDA and a requirement for marking each treated animal.

Few state agencies have understood the FDA regulatory process and the meaning of an INADA in the context of wildlife contraception, which has led to immense problems. The state usually assumes "approval" means that a drug has passed through the entire FDA drug-testing process and has been approved for sale as a commercial product. Even after they learn the nature of the INADA, which *authorizes* the use of an experimental contraceptive drug in a particular setting, the word experimental causes confusion and often concern. Beyond the state agency, county, municipality, and even park regulations must also be considered. Quite a few potential deer contraceptive projects have not materialized because of local firearms ordinances, which also apply to capture guns, or in a few cases the ordinances had to be altered for this specific purpose.

For animals living within certain federal reservations, such as national parks, national forests, wildlife refuges, Bureau of Land Management lands, and military reservations, however, there are individual and specific requirements for the testing and application of contraceptive drugs to free-ranging wildlife above and beyond those of the state, and these regulations vary from unit to unit. Managers of wildlife on federal reservations have the legal right to manage their wildlife as they see fit, without interference by state agencies. However, the federal agency may voluntarily choose to seek state cooperation and state approval. Thus, contraceptive projects on federal lands may become extremely complicated depending on the actions of their managers.

Some domestic and foreign national parks maintain ethics committees, similar to IACUCs, to review and approve, modify, or reject research proposals. For example, elephant contraceptive research in South Africa's Kruger Park required approval by both a South African National Parks Board research committee and an ethics committee. Similar contraceptive research with wild horses in two US national parks required review and approval by the Resource Management Division of the parks. White-tailed deer contraceptive research and wild horse contraceptive research in two national parks in the United States required additional approval by the regional scientist for each region.

Within national parks, the remote delivery of contraceptive drugs may only be performed by persons who have been trained in the chemical immobilization course sponsored by the US National Park Service or its equivalent. Each researcher must then be certified within the national park where the research project is occurring, by passing a delivery equipment test and receiving written certification by the superintendent of the park.

In all these federal reservations, the National Environmental Policy Act (NEPA) requires an environmental impact statement (EIS) or an environmental assessment (EA) before contraceptive drugs may be applied to free-ranging wildlife. This requirement is the responsibility of the managing agency. Within the Department of the Interior, which includes the National Park Service, the Bureau of Land Management, and the Fish and Wildlife Service, the primary research mission rests with the Biological Resources Division of the US Geological Survey. Approval for wildlife contraceptive projects may also require some level of review and approval by this group, as certainly is the case for the Bureau of Land Management in regard to wild horse contraceptive research.

Two final regulatory levels remain within the domain of free-ranging wildlife contraception, and, although neither carries the force of law, they remain the most powerful influences on whether wildlife contraception research occurs or is applied. The first of these influences originates from the research team and brings into focus ethical considerations. Often wildlife contraceptive research groups are asked to consider projects with specific species or in locations where there are clear ethical dimensions that are not resolved. Should deer be subjected to fertility control because they eat ornamental shrubbery or because a municipality will not reduce a speed limit in a high-density area? Should wolves be subjected to fertility control because they consume caribou for which a state sells hunting licenses? Should seals be subjected to fertility control because they eat food items for which commercial fishermen compete? Should an endangered species ever be contracepted? Who makes the decisions, and what are the criteria for decision making? These are but a few of the ethical issues surrounding wildlife contraception, and the research team must be prepared to accept the responsibility of an ethical evaluation of its potential actions. One serious question that all scientists must ask, and this certainly applies to those engaged in wildlife contraception, is whether the agency for which the research is being carried out plans to use the technology responsibly.

Finally, there is what we might call the "court of public opinion," which is without doubt the most powerful regulatory force of all. In the final analysis, wildlife does not belong to state agencies, or to park superintendents or to animal welfare groups, and certainly not to the scientists who pursue this seemingly bizarre approach to wildlife management. None has exclusive claim to wildlife. In the final analysis, wildlife belongs to a larger public, and their concerns and ideas cannot be ignored.

Perhaps this is nowhere a larger issue than with the concept of white-tailed deer contraception in urban areas of the United States, where it has reached epic proportions. Much of the conflict seems to be centered about the general concepts of lethal versus nonlethal approaches, although the larger issue of whether to manage at all often surfaces. Regardless of the larger public's right to have a say in the management of the wildlife it ultimately owns, a good deal of the information surrounding wildlife contraception comes to the public from highly sensationalized media hype. Nevertheless, once the public makes a decision, whether correct or incorrect, whether in the best interests of the wildlife or not, this body can create intense political pressures from the highest levels of government. Animal lovers lobby for nonlethal control, while hunters and state wildlife agencies fear a loss of hunting opportunity. Ranchers want fewer wild horses on land that they use for livestock grazing, and horse advocacy groups want more horses. Some segments of African society see elephants only as an economic commodity that can bring income from hides, meat, ivory, and hunting, but an equally large segment sees elephants as national treasures and seeks nonlethal controls. In each case the various public groups act to utilize the political process and place pressures upon state and federal legislative bodies to bring about the mode of management that each seeks.

The salient point of bringing into focus the regulatory forces of the public domain is that the scientist must navigate a careful course through these forces if the technology is ever to be applied. It is a fact that most scientists engaged in wildlife contraception start out by examining only the biological possibilities. They usually do not get very far down the research road before the legal constraints become obvious, and then, often, a good deal of research is lost because it cannot be used, perhaps because the contraceptive agent can pass through the food chain, or because it is unlikely that any oral wildlife contraceptive will ever be approved by the FDA unless it is species specific. Even if they get past the legal constraints, many scientists soon find that the weight of public opinion usually runs against highly stressful procedures, or that the expense of a particular approach is beyond the reaches of the public entities which want to use this technology, or that lack of information or even misinformation has armed the public with incorrect perceptions about the contraceptive or delivery process. In the end, public opinion is much like a mold into which the scientist must design the research and application to fit, and if the scientific community does not understand the precise shape of that mold before they start, not much will result from even the best scientific achievement.

REFERENCES

Center for Veterinary Medicine. 1994. Animal Medical Drug Use Clarification Act of 1994 (http://www.fda.gov/cvm/index/amducca/amducatoc.htm). Washington, DC: US Department of Health and Human Services, Federal Drug Administration.

Center for Veterinary Medicine. 1999. Guidance for industry: FDA approval of new animal drugs for minor uses and for minor species (http://www.fda.gov/cvm/guidance/minorgde.pdf). Washington, DC: US Department of Health and Human Services, Federal Drug Administration.

Federation of Animal Science Societies. 1999. *Guide for the Care and Use of Agricultural Animals in Agricultural Research and Teaching*. Savoy, IL: Federation of Animal Science Societies.

National Institutes of Health. 1996. *Public Health Service Policy on Humane Care and Use of Laboratory Animals*. Rockville, MD: National Institutes of Health.

National Research Council. 1996. *Guide for the Care and Use of Laboratory Animals*. Washington, DC: National Academy Press.

Part II

THE METHODS

The contraceptive methods covered in the following chapters are primarily those that are widely available and commonly used. There are many published reports of other methods that have been tested in a variety of species, but most of those either were not successful or were never actually used to manage fertility in free-ranging or captive animal populations. Thus, we are limiting coverage to those that can be obtained and administered by animal managers and veterinarians. Much of the information concerning the methods that are used in zoo mammals has been obtained from the American Zoo and Aquarium Association (AZA) Contraception Advisory Group (CAG) Contraception Database. Responses to the annual Contraception Survey, sent to AZA and many international zoos, are compiled into the Database. The results reveal what reversible contraceptive methods have been used and in which species (see Appendix). The Database is analyzed to assess efficacy and reversibility of the various contraceptive methods that, along with research results, are used to develop the CAG Recommendations (www.stlzoo.org/contraception). Promising methods that are still in development are included in the discussion on future directions in the Epilogue.

In reviewing both the literature and our Contraception Database, it is apparent that contraception has been applied to many more species and individuals in captivity than in the wild. This situation is especially true for investigations of safety and pathology following long-term use of contraceptives, because free-ranging animals are difficult if not impossible to monitor for health and cause of death and because none have been treated with contraceptives as long as have the species in captivity. This bias is reflected in the coverage in these chapters, because significantly more data on contraceptive use, rather than attempts at development, are available from captive animals.

Information on fertility control in feral domestic dogs and cats is also missing from these chapters, with few exceptions. Currently, interest is growing in identifying nonlethal and nonsurgical techniques for limiting the burgeoning populations of these feral species. However, to date, the development of these techniques is being pursued primarily in the laboratory and not in field applications. We anticipate that, given the extent of the effort being brought to bear on the problem, there will be an increase in field testing and application in the not too distant future.

CHERYL S. ASA

3
TYPES OF CONTRACEPTION
The Choices

Although the birth control pill, containing a combination of synthetic estrogen and progestin, was introduced more than 40 years ago, surprisingly few new contraceptive methods have been developed and approved since that time. Further, most of those alternatives have been little more than modifications in delivery, such as Depo-Provera injections (Pharmacia and Upjohn) and Norplant implants (Wyeth-Ayerst), which are still synthetic steroid hormones. The major differences in formulation have involved the use of newer synthetic analogues of those same steroid hormones (estrogens and progestins) and of lower doses to improve safety.

These hormones are relatively safe as well as effective in humans but, unfortunately, are not safe for certain other mammalian species. In particular, carnivore species appear to react differently. Most research has concentrated on wild felids, but considerable data are also available for canids (see Chapter 5, Adverse Effects of Contraceptives). Furthermore, for wildlife, delivery of the contraceptive is more of an issue than with humans. It is clear that a broader range of safe and effective contraceptive choices are needed for both captive and free-ranging wildlife.

The following review focuses on methods that are currently available and reported as being in use by zoo and wildlife managers. These methods are also represented in the annual recommendations published by the American Zoo and Aquarium Association (AZA) Contraception Advisory Group (found at www.stlzoo.org/contraception). Methods in earlier stages of development or in limited trials are not included here but can be found in the discussion of future directions in the Epilogue.

A PRIMER OF REPRODUCTIVE PROCESSES: POTENTIAL TARGET TISSUES OR PROCESSES FOR CONTRACEPTIVE INTERVENTION

This overview of the basic mammalian endocrine system and production of sperm and eggs is provided for the nonspecialist so that the target and mode of action of the various contraceptive techniques can be better understood.

Hormones and Reproductive Processes

The cascade of events begins, for both males and females, in the hypothalamus with the production of gonadotropin-releasing hormone (GnRH, also called luteinizing hormone-releasing hormone, LHRH), which controls release of both the gonadotropins, FSH (follicle-stimulating hormone) and LH (luteinizing hormone), from the anterior pituitary. Even though they are named for their effects on the female ovary, FSH and LH are also the pituitary hormones that support testosterone production and spermatogenesis in males (Figure 3.1).

In the ovary, the follicles, which are stimulated by FSH, secrete estradiol. Increasing blood levels of estradiol cause changes throughout the female's body, including vulvar swelling and changes in vaginal cytology and secretions and in the consistency of cervical mucus, in addition to inducing estrous behavior. When estradiol reaches a critical threshold, it prompts surges first in GnRH and then in LH, which are followed by ovulation. Estradiol may also be important for transport of both sperm and ova (eggs) to the site of fertilization. Following ovulation, the follicle switches off estradiol production and begins to secrete progesterone, which among other functions prepares the uterus for pregnancy. The ratio of estradiol to progesterone also influences embryo implantation and continued maintenance of pregnancy.

In the testes, FSH seems important for the initiation of spermatogenesis, both at puberty and at the beginning of each breeding season in species that do not produce sperm continuously. LH primarily stimulates testosterone production, which in turn stimulates and maintains spermatogenesis. As with estradiol in females, testosterone has multiple target tissues outside the testes, including those responsible for the species-specific secondary sex characteristics such as the antlers of deer, the lion's mane, and muscle development, as well as the regions of the brain that mediate aggression, territoriality, courtship, and mating.

Hypothalamic and pituitary hormones are controlled by both positive and negative feedback from the gonadal hormones (testosterone in males and estradiol and progesterone in females) that they stimulate, resulting in a tightly orchestrated

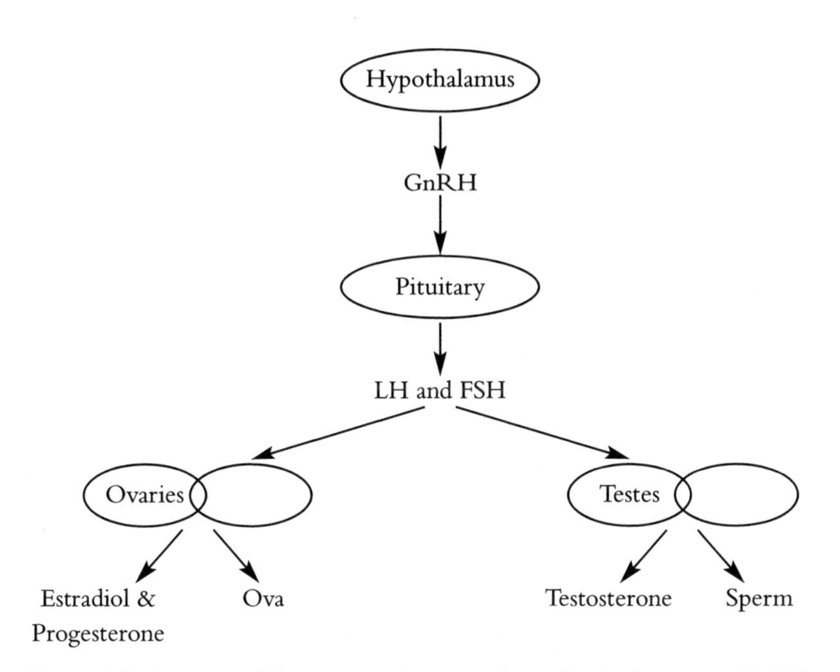

Figure 3.1. Overview of the sources and targets of reproductive hormones. *GnRH*, gonadotropin-releasing hormone; *LH*, luteinizing hormone; *FSH*, follicle-stimulating hormone.

feedback loop. The hormone GnRH from the hypothalamus stimulates the pituitary to secrete the gonadotropins LH and FSH. LH and FSH subsequently have negative feedback effects on hypothalamic release of GnRH, helping to control its secretion so that overstimulation does not result. LH and FSH have only a stimulatory effect on the ovary and testes, however, resulting in secretion of estradiol and progesterone or testosterone, which in turn feed back on both the pituitary and the hypothalamus to dampen secretion of LH, FSH, and GnRH. These negative feedback loops have been likened to the thermostat on a furnace, where a fall in temperature allows the furnace to turn on, producing heat, whereas a rise in temperature shuts down the furnace. The result is to maintain the temperature within a set, narrow range.

Thus, suppression of GnRH or the gonadotropins can interfere with gonadal hormone production as well as with follicle growth, ovulation, and spermatogenesis. In the female, progesterone and estrogen dynamics are critical for sperm and egg transport and for implantation of the embryo. Furthermore, progesterone, secreted by the corpora lutea on the ovary or by the placenta, is necessary for the support of pregnancy. Therefore, interfering with production of progesterone and estrogen may prevent implantation or the continuation of pregnancy.

Fertilization

Following ovulation, the ova travel down the oviduct to its junction with the uterus. If copulation has occurred, sperm travel through the uterus to this same junction and begin to attempt penetration of the outer protective coating of the egg, the zona pellucida (ZP). If penetration is successful, fertilization occurs; after a species-specific period of time, transport into the uterus and implantation of the fertilized egg follow.

Targets for Contraceptives

In summary, most of the currently available contraceptive methods interfere at some point in the sequence of hormone synthesis and release to control one or more reproductive events or processes (for example, ovulation, spermatogenesis, sperm or egg transport, and implantation), whereas one, the ZP vaccine, directly impedes fertilization.

PERMANENT METHODS FOR PREVENTING REPRODUCTION IN FEMALES

Gonadectomy

Removal of the ovaries eliminates the source of ova (eggs) and of the sex steroid hormones estradiol and progesterone, which precludes both ovulation and estrous behavior. Although it involves major surgery, ovariectomy (or ovariohysterectomy, which also removes the uterus; also known as spaying) may be preferable in species in which exposure to naturally occurring reproductive hormones or steroid contraceptives may be associated with serious side effects. Examples include carnivores, in which even endogenous reproductive hormones have been related to uterine infection and tumors (see Chapter 5, Adverse Effects of Contraceptives) and for individuals or species with symptoms of or a tendency to develop diabetes mellitus. There are no data available on other potential problems such as decreased bone density following removal of the ovaries in long-lived animals such as the great apes. Although the common practice in domestic dogs and cats is to remove the uterus as well as ovaries, a comparative study of the two procedures in dogs found no differences in incidence of any of the anticipated side effects (Okkens et al. 1997).

Sterilization

Tubal ligation or otherwise cutting or blocking the oviducts has not been well developed in nonhuman animals. As previously explained, gonadectomy to remove the source of reproductive steroid hormones is preferable in many instances. However, permanent surgical sterilization might in some cases be indicated, particularly in species for which gonadal hormones have not been shown to be associated with pathology. It is important in some taxa, such as canids, to ensure that the ligation is placed in the oviduct and not along the uterine horn. During estrous cycles, hormone stimulation of the uterine endometrium results in secretions that can accumulate in the horn above the stricture (Wildt and Lawler 1985). Attempts to develop techniques for reversible occlusion of the oviducts (fallopian tubes) have not yet provided reliable results.

REVERSIBLE CONTRACEPTION FOR FEMALES

Considerably more contraceptive options are available for females than for males. In general, female-directed contraception may be more straightforward because females produce fewer gametes at more discrete intervals than do males. Perhaps more importantly, however, most contraceptives have been developed for the human market in which cultural factors favor contraception for females. Only the contraceptive methods that are available for use in wildlife are covered in this chapter. Other methods that are in early stages of testing or development are presented in the discussion of future directions in the Epilogue, and methods which act solely as abortifacients are not included.

Hormonal Methods

STEROID HORMONES: PROGESTINS The female-directed contraceptives most commonly used in zoos are progestin-based products (Table 3.1). Progestins may act at any of several points in the reproductive process, including thickening cervical mucus so that sperm passage is impeded, interrupting sperm and ovum transport in the uterus and oviducts, interfering with implantation, and inhibiting the LH surge necessary for ovulation (Brache et al. 1985; Diczfalusy 1968). Typically, higher doses are needed to block ovulation than to achieve a contraceptive effect at the other points (Croxatto et al. 1982). Thus, ovulation may occur in adequately contracepted individuals (Brache et al. 1990). Progestins are not able to completely suppress follicle development, and the resulting secretion of estrogen can stimulate the physical and behavioral signs of estrus, so indications of estrus cannot be used to judge the efficacy of a progestin-based contraceptive.

Table 3.1
Synthetic progestins used as contraceptives

Synthetic progestin	Product name	Manufacturer or supplier
Altrenogest	Regu-mate oral solution	Hoechst-Roussel
Etonorgestrel	Implanon implants (Europe, Australia, Indonesia)	Organon
Levonorgestrel	Norplant implants	Wyeth-Ayerst
	Jadelle implants (Europe)	
Medroxyprogesterone acetate	Depo-Provera injections	Pharmacia and Upjohn
Megestrol acetate	Ovaban tablets	Schering-Plough
	Ovarid tablets (Europe)	Schering-Plough
Melengestrol acetate	MGA implants	E. D. Plotka
	Mazuri ADF-16 with MGA	Purina Mills LLC
	MGA200 or 500 Pre-mix	Pharmacia and Upjohn
Norethindrone	Micronor tablets	Ortho-McNeil
	NorQD tablets	Watson
Norgestrel	Ovrette tablets	Wyeth-Ayerst
Proligestone	Delvosteron injections (Europe)	Intervet

MGA, melengestrol acetate.

The basic mode of action is similar for all the synthetic progestins, via binding to progesterone receptors. However, differences occur among the progestins in the degree to which they also bind other receptors and act as other steroids, such as glucocorticoids and androgens (Duncan et al. 1964; Fekete and Szeberenyi 1965; Kloosterboer et al. 1988), and can stimulate growth hormone (Rijnberk and Mol 1997), which can result in undesirable side effects (Sloan and Oliver 1975; Selman et al. 1997). In fact, melengestrol acetate (MGA, available from E. D. Plotka) was chosen for the implants now commonly used in zoos, because it stimulated glucocorticoid release less than did medroxyprogesterone acetate (MPA, the synthetic progestin in Depo-Provera), the other progestin widely available at the time (Seal et al. 1976). A further problem with MPA is its documented androgenic activity, equated in some tests with dihydrotestosterone (Labrie et al. 1987), a natural androgen with potent morphological effects, especially during development. Among currently available progestins, levonorgestrel (the progestin in Norplant) has the highest binding affinity to androgen receptors, which is considered a potential health risk in humans because of its effect on lipid profiles and the cardiovascular system (Sitruk-Ware 2000).

In some species, progestin supplementation (Diskin and Niswender 1989) may help maintain pregnancy whereas, in others, especially early in gestation, this practice has been associated with embryonic resorption (Shirley et al. 1995; Ballou 1996). Later in pregnancy, progestins might be predicted to interfere with parturition because they are known to suppress contractility of uterine smooth muscle. Such an effect was documented in white-tailed deer (Plotka and Seal 1989), but females of some other species treated with progestins have given birth without incident (Porton 1995; see Chapter 8, Contraception in Nonhuman Primates). The discrepancy may be a result of dosage (the dose required for contraceptive efficacy may be lower than that which inhibits parturition), the progestin (some types may have more effect on the uterine myometrium or be more bioactive), and species differences (Zimbelman et al. 1970; Jarosz and Dukelow 1975; Plotka and Seal 1989; Shirley et al. 1995).

In contrast to the variability in effect on pregnancy, progestins appear to be generally safe for lactating females and nursing young. They do not interfere with milk production and have not been found to have negative effects on the growth or development of nursing infants (WHO 1994a, 1994b).

MGA, the contraceptive most widely used in US zoos for the past 25 years, has been effective in virtually all mammalian taxa in which it has been tested (see Chapters 7 through 11). Developed by Dr. U. S. Seal in the mid-1970s, the Silastic implants containing MGA are now supplied without cost to the zoo community by Dr. E. D. Plotka. More recently, a new formulation was introduced for use in hoofstock that incorporates MGA into one of the most commonly used commercial hoofstock diets (Mazuri; Purina Mills, LLC). A disadvantage of this approach is confirming that each animal in the group consumes the needed dose each day. In the planning stages is another formulation of MGA for hoofstock that would be given daily to individual animals in their favorite treat to ensure ingestion.

Other progestins frequently used in captive animals include medroxyprogesterone acetate (Depo-Provera), levonorgestrel (Norplant), and megestrol acetate (Ovaban, Ovarid). Progestins have proven very effective in most mammalian species, approaching 100 percent efficacy when administered properly. Equids are the most notable exception in the list of species successfully treated with the more common synthetic progestins. Only altrenogest (Regu-mate: Hoechst-Roussel), another synthetic progestin that is effective in domestic horses for synchronizing estrus (Webel and Squires 1982), should also be effective as a contraceptive but at a higher dose. A drawback for application to equids is that it is only available for oral administration, but that feature is considered an advantage when treating marine mammals (see Chapter 10, Contraception in Pinnipeds and Cetaceans).

STEROID HORMONES: ESTROGENS Estrogens can effectively suppress follicle growth so that ovulation is prevented, but at contraceptive doses these have been associated with serious side effects in many species (Dunn and Green 1963; Gass et al. 1964). The estrogens diethylstilbestrol (DES), mestranol, estradiol benzoate, and estradiol cypionate have been used to block implantation following mismating in dogs. However, the tendency of estrogens to stimulate uterine disease, bone marrow suppression, aplastic anemia, and ovarian tumors makes them inappropriate contraceptive compounds (Bowen et al. 1985).

STEROID HORMONES: ESTROGEN–PROGESTIN COMBINATIONS

Some of the side effects associated with estrogen treatment, for example, overstimulation of the uterine lining in primates, can be mitigated by adding a progestin. However, the addition of progestin to estrogen is synergistic, not inhibitory, in carnivores, so the combination is even more likely to result in uterine and mammary disease (Brodney and Fidler 1966; reviewed in Asa and Porton 1991). Because this synergy occurs in canids when progestin-only methods are initiated during proestrus, when natural estrogens are elevated, treatment should be initiated well in advance of the breeding season. When treatment is begun during deep anestrus, the side effects of synthetic progestins are minimized (Bryan 1973), even when continued for several years, a regimen that has proven safe in domestic dogs during several decades of use in Europe (W. Jöchle, personal communication).

More than 50 orally active contraceptive products containing various combinations of an estrogen and a progestin at various doses are currently approved for human use in the United States (PDR 2002). Ethinyl estradiol is the most common form of estrogen (53 products), although a few (4) contain mestranol. Norethindrone is the most common progestin ingredient; other combinations include levonorgestrel, desogestrel, noregestrel, norgestimate, and ethynodiol diacetate. Oral contraceptive regimens designed for humans are intended to simulate the 28-day menstrual cycle, with 21 days of treatment followed by 7 days when either a placebo or no pill is taken, resulting in withdrawal bleeding that resembles menstruation. When the pill was first developed, it was thought that women preferred to have regular menstrual periods that assured them their reproductive processes were normal (Pincus 1965). It was also feared that continued treatment with the combination of estrogen and progesterone might overstimulate the uterine endometrium. In fact, continual delivery of ethinyl estradiol and melengestrol acetate from Silastic implants in spider monkeys (*Ateles geoffroyi*) resulted in endometrial growth, detected by uterine biopsy, that prompted cessation of the study after only a few months (C. Asa and I. Porton, unpublished observations). However, recent data from women indicate that continuous use of combination pills for up to 6 months (length of the study) results in an inactive endometrium (Kwiecien et al. 2003).

Another contraceptive formulation combining an estrogen and progestin approved more recently was the monthly injection Lunelle, which contains MPA and estradiol cypionate (Pharmacia and Upjohn). However, it was withdrawn from the market in late 2002.

STEROID HORMONES: ANDROGENS Both testosterone and the synthetic androgen mibolerone (Cheque Drops: Pharmacia and Upjohn) have proven effective contraceptives (domestic dog: Simmons and Hamner 1973; Sokolowski and Geng 1977; domestic cat: Burke et al. 1977; and gray wolf, *Canis lupus*; leopard, *Panthera pardus*; jaguar, *P. onca*; lion, *P. leo*: Gardner et al. 1985). However, masculinizing effects included clitoral hypertrophy, vulval discharge, mane growth (female lion), mounting, and increased aggression. Mibolerone is approved for administration to dogs but is contraindicated for females that have impaired liver function or are lactating or pregnant because female fetuses can become virilized. Although mibolerone has been tested and found effective, it is not approved for use in cats. Side effects, which appear more serious than those in dogs, include changes in thyroid function, increased serum cholesterol, and possible uterine disease (Burke et al. 1977). Especially because of the potential for increased aggression, mibolerone use in wildlife is inadvisable.

GONADOTROPIN-RELEASING HORMONE ANALOGUES Synthetic agonists of GnRH (gonadotropin-releasing hormone from the hypothalamus) have a higher affinity for GnRH receptors and a longer half-life in the circulation, which makes them more effective at lower doses and for longer periods of time than natural GnRH. Synthetic forms can be *antagonists*, that block the action of the natural hormone, or *agonists*, which have the same effects (in this case, stimulatory) on target tissue as the natural hormone. Although antagonists would be assumed the more logical selection for contraception, they are considerably more expensive, shorter acting, and less safe than the agonists, which limits their application (Vickery et al. 1989). Two GnRH antagonists are currently approved for human use for short-term suppression of ovulation during assisted fertility protocols: cetrorelix (Cetrotide: Serono) and ganirelix (Antagon: Organon) (Table 3.2). These injectable formulations are active for 1 to 4 days, making them unsuitable for contraception.

In contrast to antagonists, GnRH agonist administration is followed by two distinct response phases. In the acute phase, lasting for several days, both LH and FSH

Table 3.2
Currently available gonadotropin-releasing hormone (GnRH) agonists and antagonists

Product name	Generic name	Manufacturer or supplier
Long-acting agonists:		
Decapeptyl Depot	Triptorelin acetate (Europe)	Ferring
Lupron Depot injection	Leuprolide acetate	TAP Pharmaceuticals
Profact Depot injection	Buserelin (Europe)	Aventis
Suprelorin implant	Deslorelin	Peptech Animal Health
Synarel nasal spray	Nafarelin	Searle
Viadur Implant	Leuprolide acatate	Bayer
Zoladex implant	Goserelin	Astra Zeneca
Antagonists:		
Antagon	Ganirelix	Organon
Cetrotide	Cetrorelix	Serono

are stimulated, which can result in estrus and ovulation if follicles of the right stage are present (Bergfield et al. 1996; Maclellan et al. 1997). Continued treatment with long-acting preparations, such as implants or microspheres, is associated with the chronic phase during which secretions of both FSH and pulsatile LH are blocked due to downregulation of GnRH receptors in the cells that produce LH and FSH (Huckle and Conn 1988). GnRH agonists also can interfere directly with follicle development (Parborell et al. 2002). The observed effects in the animal are similar to those following ovariectomy but are reversed after the hormone content of the implant or microspheres is depleted.

A method for preventing estrus and ovulation during the acute phase has been tested in domestic dogs. The synthetic oral progestin megestrol acetate (Ovaban or Ovarid) given just before and during the first week following implant insertion successfully prevented both proestrus and estrus. However, even the females that are not progestin treated and which conceive during the acute phase do not successfully maintain pregnancy (Wright et al. 2001).

Five long-acting GnRH agonists are currently available in the United States (see Table 3.2): leuprolide, as Lupron Depot (TAP Pharmaceuticals), in injectable formulations lasting 1, 3, or 4 months, and Viadur Implant (Bayer), a nonbiodegradable implant with an osmotic pump effective for 1 year (both products are approved only for prostate cancer treatment); goserelin (Zoladex: Astra Zeneca), a biodegradable implant effective for 3 months; nafarelin acetate (Synarel: Searle), as a nasal spray, approved for treatment of precocious puberty and endometriosis; and deslorelin (Suprelorin: Peptech Animal Health, Australia, available in the United States by arrangement with the AZA Wildlife Contraception Center at the Saint Louis Zoo), a biodegradable implant in two sizes with durations of efficacy from about 6 months to 2 years, depending on implant size and species treated. Several other formulations are available in Europe (see Table 3.2). A short-acting deslorelin implant (Ovuplant: Peptech Animal Health), designed to induce ovulation in horses, is not suitable as a contraceptive.

Deslorelin has been effective in female domestic dogs (Trigg et al. 2001), domestic cats (Munson et al. 2001), domestic cows (D'Occhio et al. 2000), and several exotic species (lion; leopard; cheetah, *Acinonyx jubatus*; fennec fox, *Vulpes zerda*; wild dog, *Lycaon pictus*: Bertschinger et al. 2001, 2002). Fertility was restored when the implants were removed or their hormone content was depleted; no pathology has been reported.

Immunocontraception: Zona Pellucida Vaccines

One of the early stages of fertilization involves the binding of sperm to the zona pellucida (ZP), the glycoprotein coating of the mammalian oocyte, or egg. Immunization with ZP proteins results in antibodies that reversibly interfere with this process. Initial treatment requires at least two injections, about 1 month apart, with subsequent boosters needed annually for seasonal breeders but perhaps more frequently for continuous breeders.

When the ZP antigen derived from porcine ova (PZP) is injected intramuscularly into females of species other than pigs, the target animal's immune system is stimulated to produce antibodies against the antigen. These antibodies attach to the sperm receptors of ova and cause steric hindrance, which then blocks fertilization (Paterson and Aitken 1990). Porcine ZP has been effective in a wide variety of ungulates and some carnivores (Kirkpatrick et al. 1996), is safe when administered during pregnancy or lactation, and is reversible after short-term use. However, long-term studies with white-tailed deer (*Odocoileus virginianus*) and feral horses (*Equus caballus*) revealed that treatment for 5 or more years was increasingly associated with ovarian failure (Kirkpatrick et al. 1997). The possibility for permanent ovarian damage makes this method unsuitable for animals valuable to captive breeding programs or other cases in which reversibility is important.

When ZP vaccination results in permanent sterilization, it appears to act not only on the ZP itself but also on the oocytes or their surrounding granulosa cells (VandeVoort et al. 1995). The type and extent of the response appears to depend primarily on two factors: (1) the degree of similarity between the species in which the antibody is raised and the species being treated; and (2) the stages of ovarian follicular development, that is, the absence of late-stage developing follicles prevents recruitment of oocytes that are in meiotic arrest so that they are destroyed by the ZP antigens until all are depleted (Dunbar et al. 2002).

When vaccine effect is restricted to preventing sperm entry so that ovarian activity is not disrupted, ovulatory cycles with estrous behavior continue. In some species, for example, white-tailed deer, the failure to conceive results in a longer than usual breeding season, with continued estrous cycles accompanied by rutting behavior (McShea et al. 1997). Continued breeding activity may be desirable in some situations where it is seen as more natural than suppression, but it can also result in aggression and social disruption, especially when pregnancy does not ensue.

A further problem with ZP vaccines is that they are most effective when administered with Freund's complete adjuvant, which can produce reactions at the injection site. In some species, such as felids, a more serious reaction may result in systemic pathology. Other adjuvants are being developed and tested.

Mechanical Methods: Intrauterine Devices

Intrauterine devices (IUDs) prevent pregnancy primarily by local mechanical effects on the uterus that impede implantation. Most designs include an electrolytic copper coating, which increases efficacy, because the copper ions are spermicidal. Although IUDs have been associated with pelvic inflammatory disease in humans, the risk of infection is statistically increased only during the first 4 months after insertion (Lee et al. 1983). Monofilament tail strings do not increase that risk (Triman and Liskin 1988), but, rather, attention to aseptic technique during insertion, with or without prophylactic antibiotics, is critical to preventing infection. In humans, the chance for infection is higher for women who have a history of tubal infection, are nulliparous and under 25 years of age, or who have multiple sexual partners (Lee et al. 1988; Daling et al. 1985; Cramer et al. 1985). However, the IUD is considered safe and very effective when used by parous females in monogamous relationships (Chi 1993). In fact, it can be ideal for lactating females (Díaz et al. 1997).

The IUDs marketed for humans (Table 3.3) may be appropriate for species with a uterine size and shape comparable to that of humans, in particular, the great apes (for example, orangutan, *Pongo pygmaeus*: Florence et al. 1977; chimpanzee, *Pan troglodytes*: Gould and Johnson-Ward 2000) or even other primates (for example, rhesus, *Macaca mulatta*: Mastrioanni et al. 1967). Various IUDs tested in domestic cows (Turin et al. 1997), ewes (Ginther et al. 1966), and goats (Gadgil et al. 1968) have also been shown to be effective, although in some cases estrous

Table 3.3
Currently available intrauterine devices

Product name	Composition	Manufacturer
Mirena	Polyethylene T encased in Silastic that releases levonorgestrel	Berlex
ParaGard T 380A	Polyethylene T wound with copper wire	Ortho-McNeil

cycles have been suppressed, calling into question their site of action. An IUD recently developed for domestic dogs (Biotumer, Argentina, SA) has been found safe and effective in limited trials (Nagle and Turin 1997; Volpe et al. 2001). Its flexible Y-shape allows it to lie within the uterine body with each arm in a uterine horn. The catheter with cervical distender and application tube aids insertion. There are two sizes, for dogs of more than or less than 12 kg.

EFFECTS ON BEHAVIOR

Despite the decades of contraceptive use in wildlife, few studies have focused on associated behavior. The overwhelming concern has been on efficacy and safety, which is of course appropriate during the initial stages of development and application. Now that contraception is so widely used in so many species, however, it is surprising that so little attention has been directed toward possible effects on behavior or social groups. The most obvious effect of ovariectomy and GnRH agonists is the elimination of sexual activity. Progestins may also suppress estrus, but typically only at higher doses; progestin–estrogen combinations are more likely to inhibit the follicular growth associated with estrous behavior, which itself can have further implications for social interaction (see Chapter 6, Choosing the Most Appropriate Contraceptive). IUDs, and in many cases PZP vaccines, do not affect estrous cycles.

Beyond obvious effects on estrous behavior, research has linked progestin use with mood changes (MPA: Sherwin and Gelfun 1989), depression (MPA: Civic et al. 2000), and lethargy (Evans and Sutton 1989). However, studies of social groups of hamadryas baboons (*Papio hamadryas*: Portugal and Asa 1995), Rodrigues fruit bats (*Pteropus rodricensis*: Hayes et al. 1996), golden lion tamarins (*Leontopithecus rosalia*: Ballou 1996), golden-headed lion tamarins (*Leontopithecus chrysomelas*: De Vleeschouwer et al. 2000), and lions (Orford 1996) found no significant effects on behavior or interactions of group members despite treatment of some or all females with MGA. Although MPA (Depo-Provera) was associated with increased aggressive behavior in female stumptail macaques (*Macaca arctoides*: Linn and Steklis 1990), feral domestic cats treated with megestrol acetate (Ovaban and Ovarid) were described as more docile (Remfry 1978). This difference may result from the higher affinity of MPA for androgen receptors (Labrie et al. 1987), so that MPA may act on behavior more as an androgen than a progestin.

PERMANENT METHODS FOR PREVENTING REPRODUCTION IN MALES

Gonadectomy

Male castration is a simple procedure in most species but may not be practical in those with undescended or partially descended testes (for example, pinnipeds, cetaceans, elephants). However, effects on secondary sex characteristics caused by the decline in testosterone may involve complete loss (for example, the lion's mane) or disruption of the seasonal cycle (for example, deer antlers that stay in velvet). Following castration, especially in sexually experienced males, libido may decline slowly if at all, but males may not be able to copulate successfully, usually because of a failure to achieve erection. Declining testosterone levels following castration may result in reduced aggression, but learned behavior patterns may persist.

Sterilization

Vasectomy is an option for males in which maintenance of secondary sex characteristics and male-type behavior is desirable. Although potentially reversible, the technique requires highly skilled microsurgery that has a variable success rate. Antisperm antibodies detected in men following vasectomy were expected to limit the success of vasectomy reversal (Gupta et al. 1975; Alexander 1977). However, high pregnancy rates have been achieved by some practitioners (Silber 1989a). Of 282 patients who had sperm in the vas before reversal surgery (vasovasostomy), 81 percent successfully reversed, as evidenced by pregnancies in the partners. Even in some cases where sperm were not present in the vas, microsurgical connection of the vas to the epididymis (vasoepididymostomy) was often successful (Silber 1989b). The success rate (pregnancies in partners) was higher for anastomosis involving the body (56 percent) than the head (31 percent) of the epididymis, probably because those sperm in the body were more mature.

The success rate for vasectomy reversals can be improved if the vasectomy itself is done with reversal in mind. One of the primary reasons for permanent damage is related to the pressure increase in the epididymis and testis following vas obstruction. Hence, a technique that leaves the testis end of the vas open lessens the chance of pressure-related damage and thus can increase the likelihood of successful reversal (Silber 1979; Shapiro and Silber 1979).

As an alternative to surgical vasectomy, permanent obstruction of sperm passage also can be accomplished by injecting a sclerosing agent into the cauda epididymis or vas deferens (Freeman and Coffey 1973; Pineda and Dooley 1984; Pineda et al. 1977). Treatment of the epididymis is probably more successful, because the lumen of the tubule is likely crossed several times during injection, but this technique must be considered irreversible. Treatment of a discrete area of the vas would be more amenable to reversal, by excision and reanastomosis, but might not be as effective for ensuring sperm blockage.

Vasectomy is contraindicated in felids and other carnivores that have induced ovulation followed by pseudopregnancy because those periods of prolonged, elevated progesterone can contribute to uterine or mammary gland pathology in the partner. Even in canids, the obligate pseudopregnancy with elevated progesterone following spontaneous ovulation may also contribute to uterine pathology. If so, then any method that allows repeated nonfertile cycles without intervening pregnancies should be avoided. This category would include simple separation of males from females, as well as treatments that render the male sterile.

REVERSIBLE CONTRACEPTION FOR MALES

The only commercially available contraceptive for human males is the condom, which for obvious reasons means that there are no methods that can be used off-label for wildlife. Explanations for this lack of alternatives include the higher motivation of women to control conception and the absence of a natural model of male infertility to correspond with the periods of pregnancy, lactation, and menopause in women (Silvestrini 2002).

Hormonal Methods: Gonadotropin-Releasing Hormone Agonists

The basic mechanism of action of GnRH agonists on LH and FSH in males is similar to that in females (see foregoing), with an initial increase in testosterone followed by chronic suppression. In domestic dogs, deslorelin can achieve azoospermia (absence of sperm), probably as a result of testosterone suppression (Trigg et al. 2001). Testosterone, testis size, and sperm production were suppressed in cheetahs and a wild dog treated with deslorelin. Trials with other wild canid males have been less successful (gray wolf; red wolf, *Canis rufus*; bush dog, *Speothos venaticus*: Bertschinger et al. 2001, 2002) but the experiments will be repeated at higher doses and more in advance of the breeding season.

GnRH agonists, even at extremely high doses, have not been effective in blocking either testosterone or spermatogenesis in domestic cattle (D'Occhio and Aspden 1996), horses (Brinsko et al. 1998), or the other artiodactyls in which it has been evaluated (red deer, *Cervus elaphus*: Lincoln 1987; zebu, *Bos indicus*: D'Occhio and Aspden 1996; gerenuk, *Litocranius walleri*: Penfold et al. 2002). In these species, GnRH agonists succeed in blocking the pulsatile but not basal secretion of both LH and testosterone (D'Occhio and Aspden 1996). This level of testosterone is sufficient for supporting both spermatogenesis and male behavior.

Lupron-Depot has been used successfully in a variety of species, but primarily in male marine mammals. Because of the continuous nature of spermatogenesis, the GnRH antagonists that are currently commercially available are much too short acting (see Reversible Contraception for Females, earlier in this chapter).

As cautioned earlier in the section on Sterilization, because GnRH analogues administered to males render them infertile, such treatments may not be appropriate for carnivores because of the possible risk to their female partners of uterine or mammary pathology consequent to repeated nonfertile cycles accompanied by periods of elevated progesterone.

Effects on Behavior

To the extent that GnRH agonists succeed in suppressing testosterone, the effects on behavior should be similar to those following castration. In fact, GnRH agonists have been used in males for both contraception and aggression control (see Chapters 7 through 11 on application by taxon and Chapter 12 on aggression control).

Delivery Methods

Delivery methods for the contraceptives currently available for wildlife include implants, injections, and pills. The most obvious advantage to implants is the relatively long period of hormone delivery per handling episode. Steroids are most amenable to this route of administration because they diffuse readily from Silastic. However, newer implant matrices are required for release of peptides such as GnRH. For example, the deslorelin implant consists of a matrix of low melting point lipids and a biological surfactant (Trigg et al. 2001).

Problems with implants include possible loss and migration or fragility (for example, deslorelin implants). Loss can be minimized by ensuring that sterile technique is used during insertion, and MGA implants should be gas-sterilized before insertion because local infection can result in implant loss (deslorelin and commercially available implants are presterilized). For social species, when a surgical incision is required for implant insertion, the individual should be separated from the group to prevent grooming until the incision is healed. Smaller implants such as Norplant and deslorelin may be less prone to loss for reason of their size and because they are inserted by trocar (a large needle), which reduces chances of loss by grooming as there is no incision site. Loss of subcutaneously placed Silastic implants may be common in perissodactyls (Plotka et al. 1988), and this problem requires further study.

Adding radiopaque material or an identity transponder microchip to MGA implants facilitates confirming presence and monitoring position. MGA implants can also be sutured to the muscle to impede migration. However, these modifications are not recommended for implants made from Silastic tubing (for example, Norplant) and are not possible for solid implants (for example, deslorelin), because hormone release rates may be altered.

Injectable depot preparations have been formulated to release either peptide or steroid hormones. For example, a single injection may be effective from 1 to 4 months (Lupron-Depot: 1, 3, or 4 months; Depo-Provera: 2 to 3 months). Vaccines also are administered by injection. Although remote delivery via dart is possible for injectables, delivery of the complete dose cannot always be ensured or confirmed. A further problem with vaccines that use Freund's complete adjuvant (FCA) is that the adjuvant itself can cause a reaction at the injection site or even result in systemic pathology in some species (see Chapter 5, Adverse Effects of Contraceptives). In addition, FCA administration results in treated animals subsequently testing positive for tuberculosis, which may preclude its use in some species.

A disadvantage of biodegradable implants such as deslorelin is that, unlike Norplant and MGA implants that can be removed rather easily, they are somewhat fragile and prone to breakage when handled. Coupled with the variable duration of efficacy by species and by individual that has been reported, the inability to remove the deslorelin implants is a considerable disadvantage. Nor can reversal time be controlled with depot injections and vaccines, primarily because the duration of efficacy differs markedly among individuals. However, ease of application of injectable products and the safety of GnRH agonists may be more important than timed reversals in some circumstances.

Oral delivery can be relatively simple in mammals that have been trained for daily contact or handling, such as marine mammals used in shows. However, a disadvantage of oral preparations is that they typically must be administered daily, although usually these can be incorporated into food. Confirmation of ingestion is critical and can be difficult, especially in great apes.

Contraceptive delivery presents a more serious challenge for free-ranging than for captive wildlife. Locating, baiting, and capturing can be difficult, costly, and time consuming. In fact, as much if not more effort may have been expended in developing and testing delivery techniques than on the contraceptives themselves.

Various types of dart guns and blowpipes have been used, most commonly to administer fluid-filled darts or solid matrix "biobullets" (Kreeger 1997). Although some contraceptives require capture and handling of individual animals (for example, inserting implants: Bertschinger et al. 2001; Plotka et al. 1988), such an approach is not practical for most applications. Baiting may be the most problematic method, because it must be sufficiently attractive to the target species to lure enough animals to have a population effect without attracting nontarget species. Additionally, controlling the dose ingested by each individual can be difficult, if not impossible.

ACKNOWLEDGMENTS

The author thanks Dr. Wolfgang Jöchle for suggestions on the manuscript.

REFERENCES

Alexander, N. 1977. Vasectomy and vasovasostomy in rhesus monkeys: the effect of circulating antisperm antibodies on fertility. *Fertil Steril* 28:562–569.

Asa, C. S., and I. Porton. 1991. Concerns and prospects for contraception in carnivores. In *Proceedings, American Association of Zoo Veterinarians*, 298–303, September 28–October 3, 1991, Calgary, Alberta, Canada.

Ballou, J.D. 1996. Small population management: contraception of golden lion tamarins. In *Contraception in Wildlife*, Book 1, ed. P. N. Cohn, E. D. Plotka, and U. S. Seal, 339–358. Lewiston, NY: Edwin Mellen.

Bergfield, E. G. M., M. J. D'Occhio, and J. E. Kinder. 1996. Pituitary function, ovarian follicular growth, and plasma concentrations of 17β-estradiol and progesterone in prepubertal heifers during and after treatment with the luteinizing hormone-releasing hormone agonist deslorelin. *Biol Reprod* 54:776–782.

Bertschinger, H. J., C. S. Asa, P. P. Calle, J. A. Long, K. Bauman, K. DeMatteo, W. Jöchle, T. E. Trigg, and A. Human. 2001. Control of reproduction and sex related behaviour in exotic wild carnivores with the GnRH analogue deslorelin. *J Reprod Fertil Suppl* 57:275–283.

Bertschinger, H. J., T. E. Trigg, W. Jöchle, and A. Human. 2002. Induction of contraception in some African wild carnivores by downregulation of LH and FSH secretion using the GnRH analogue deslorelin. *Reproduction Suppl* 60:41–52.

Bowen, R. A., P. N. Olson, and M. D. Behrendt. 1985. Efficacy and toxicity of estrogens commonly used to terminate canine pregnancy. *J Am Vet Med Assoc* 186:783–788.

Brache, V., A. Faundes, and E. Johansson. 1985. Anovulation, inadequate luteal phase, and poor sperm penetration in cervical mucus during prolonged use of Norplant implants. *Contraception* 31:261–273.

Brache, V., F. Alvarez-Sanchez, A. Faundes, A. S. Tejada, and L. Cochon. 1990. Ovarian endocrine function through five years of continuous treatment with Norplant® subdermal contraceptive implants. *Contraception* 41:169–177.

Brinsko, S. P., E. L. Squires, B. Pickett, and T. M. Nett. 1998. Gonadal and pituitary responsiveness of stallions is not down-regulated by prolonged pulsatile administration of GnRH. *J Androl* 19:100–109.

Brodney, R. S., and I. J. Fidler. 1966. Clinical and pathological findings in bitches treated with progestational compounds. *J Am Vet Med Assoc* 149:1406–1415.

Bryan, H. S. 1973. Parenteral use of medroxyprogesterone acetate as an antifertility agent in the bitch. *Am J Vet Res* 34:659–663.

Burke, T. J., H. A. Reynolds, and J. H. Sokolowski. 1977. A 180-day tolerance-efficacy study with mibolerone for suppression of estrus in the cat. *Am J Vet Res* 38:469–476.

Chi, I. 1993. What we have learned from recent IUD studies: a researcher's perspective. *Contraception* 48:81–108.

Civic, D., D. Scholes, L. Ichikawa, A. Z. LaCroix, C. K. Yoshida, S. M. Ott, and W. E. Barlow. 2000. Depressive symptoms in users and non-users of depot medroxyprogesterone acetate. *Contraception* 61:385–390.

Cramer, D. W., I. Schiff, S. C. Schoenbaum, M. Gibson, S. Belisle, B. Albrecht, R. J. Stillman, M. J. Berger, E. Wilson, B. V. Stadel, and M. Seibel. 1985. Tubal infertility and the intrauterine device. *N Engl J Med* 312:941–947.

Croxatto, H., S. Díaz, M. Pavez, P. Miranda, and A. Brandeis. 1982. Plasma progesterone levels during long-term treatment with levonorgestrel silastic implants. *Acta Endocrinol* 101:307–311.

Daling, J. R., N. S. Weiss, B. J. Metch, W. H. Chow, P. M. Soderstrom, D. E. Moore, L. R. Spadoni, and B. V. Stadel. 1985. Primary tubal infertility in relation to the intrauterine device. *N Engl J Med* 312:937–941.

De Vleeschouwer, K., L. Van Elsacker, M. Heistermann, and K. Leus. 2000. An evaluation of the suitability of contraceptive methods in golden-headed lion tamarins (*Leontopithecus chrysomelas*), with emphasis on melengestrol acetate (MGA) implants. (II) Endocrinological and behavioural effects. *Anim Welf* 9:385–401.

Díaz, S., A. Zepeda, X. Maturana, M. V. Reyes, P. Miranda, M. E. Casado, O. Peralto, and H. B. Croxatto. 1997. Fertility regulation in nursing women. *Contraception* 56: 223–232.

Diczfalusy, E. 1968. Mode of action of contraceptive drugs. *Am J Obstet Gynecol* 100: 136–163.

Diskin, M. G., and G. D. Niswender. 1989. Effect of progesterone supplementation on pregnancy and embryo survival in ewes. *J Am Sci* 67:1559–1563.

D'Occhio, M. J., and W. J. Aspden. 1996. Characteristics of luteinizing hormone (LH) and testosterone secretion, pituitary responses to LH-releasing hormone (LHRH) and reproductive function in young bulls receiving the LHRH agonist deslorelin: effect of castration on LH responses to LHRH. *Biol Reprod* 54:45–52.

D'Occhio, M. J., G. Fordyce, T. R. White, W. J. Aspden, and T. E. Trigg. 2000. Reproductive responses of cattle to GnRH agonists. *Anim Reprod Sci* 60-61:433–442.

Dunbar, B. S., G. Kaul, M. Prasad, and S. M. Skinner. 2002. Molecular approaches for the evaluation of immune responses to zona pellucida (ZP) and development of second-generation ZP vaccines. *Reproduction Suppl* 60:9–18.

Duncan, G. L., S. C. Lyster, J. W. Hendrix, J. J. Clark, and H. D. Webster. 1964. Biologic effects of melengestrol acetate. *Fertil Steril* 15:419–432.

Dunn, T. B., and A. W. Green. 1963. Cysts of the epididymis, cancer of the cervix, granular cell myoblastoma, and other lesions after estrogen injection in newborn mice. *J Natl Cancer Inst* 31:425–498.

Evans, J. M., and D. J. Sutton. 1989. The use of hormones, especially progestagens, to control oestrus in bitches. *J Reprod Fertil Suppl* 39:163–173.

Fekete, G., and S. Szeberényi. 1965. Data on the mechanism of adrenal suppression by medroxyprogesterone acetate. *Steroids* 6:159–166.

Florence, B. D., P. J. Taylor, and T. M. Busheikin. 1977. Contraception for a female Borneo orangutan. *J Am Vet Med Assoc* 171:974–975.

Freeman, C., and D. S. Coffey. 1973. Sterility in male animals induced by injection of chemical agents into the vas deferens. *Fertil Steril* 24:884–890.

Gadgil, B. A., W. E. Collins, and N. C. Buch. 1968. Effects of intrauterine spirals on reproduction in goats. *Indian J Exp Biol* 6:138–140.

Gardner, H. M., W. D. Hueston, and E. F. Donovan. 1985. Use of mibolerone in wolves and in three *Panthera* species. *J Am Vet Med Assoc* 187:1193–1194.

Gass, G. H., D. Coats, and N. Graham. 1964. Carcinogenic dose-response curve to oral diethylstibestrol. *J Natl Cancer Inst* 33:971–977.

Ginther, O. J., A. L. Pope, and L. E. Casida. 1966. Local effect of an intrauterine plastic coil on the corpus luteum of the ewe. *J Anim Sci* 25:472–475.

Gould, K. G., and J. Johnson-Ward. 2000. Use of intrauterine devices (IUDs) for contraception in the common chimpanzee (*Pan troglodytes*). *J Med Primatol* 29:63–69.

Gupta, I., S. Dhawan, G. D. Goel, and K. Saha. 1975. Low fertility rate in vasovasostomized males and its possible immunologic mechanism. *Int J Fertil* 20:183–191.

Hayes, K. T., A. T. C. Feistner, and E. C. Halliwell. 1996. The effect of contraceptive implants on the behavior of female Rodrigues fruit bats, *Pteropus rodricensis*. *Zoo Biol* 15:21–36.

Huckle, W. R., and P. M. Conn. 1988. Molecular mechanism of gonadotropin-releasing hormone action: I. The GnRH receptor. *Endocr Rev* 9:379–386.

Jarosz, S. J., and W. R. Dukelow. 1975. Effect of progesterone and medroxyprogesterone acetate on pregnancy length. *Lab Anim Sci* 35:156–158.

Kirkpatrick, J. F., P. P. Calle, P. Kalk, I. K. M. Liu, and J. W. Turner. 1996. Immunocontraception of captive exotic species. II. Formosan sika deer (*Cervus nippon taiouwanus*), axis deer (*Cervus axis*), Himalayan tahr (*Hemitragus jemlahicus*), Roosevelt elk (*Cervus elaphus roosevelti*), Reeves' muntjac (*Muntiacus reevesi*) and sambar deer (*Cervus unicolor*). *J Zoo Wildl Med* 27:482–495.

Kirkpatrick, J. F., J. W. Turner Jr., I. K. M. Liu, R. Fayrer-Hosken, and A. T. Rutberg. 1997. Case studies in wildlife immunocontraception: wild and feral equids and white-tailed deer. *Reprod Fertil Dev* 9:105–110.

Kloosterboer, H. J., C. A. Vonk-Noordegraff, and E. W. Turpijn. 1988. Selectivity in progesterone and androgen receptor binding of progestagens used in oral contraceptives. *Contraception* 38:325–332.

Kreeger, T. J. 1997. Overview of delivery systems for the administration of contraceptives to wildlife. In *Contraception in Wildlife Management*, ed. T. J. Kreeger, 29–48. Technical Bulletin No. 1853. Washington, DC: US Dept. of Agriculture, Animal and Plant Health Inspection Service.

Kwiecien, M. A. Edelman, M. D. Nichols, and J. T. Jensen. 2003. Bleeding patterns and patient acceptability of standard or continuous dosing regimens of a low-dose oral contraceptive: a randomized trial. *Contraception* 67:9–13.

Labrie, C., L. Cusan, M. Plante, S. Lapointe, and F. Labrie. 1987. Analysis of the androgenic activity of synthetic "progestins" currently used for the treatment of prostate cancer. *J Steroid Biochem* 28:379–384.

Lee, N. C., G. L. Rubin, H. W. Ory, and R. T. Burkman. 1983. Type of intrauterine device and the risk of pelvic inflammatory disease. *Obstet Gynecol* 62:1–6.

Lee, N. C., G. L. Rubin, and R. Borucki. 1988. The intrauterine device and pelvic inflammatory disease revisited: new results from the Women's Health Study. *Obstet Gynecol* 72:1–6.

Lincoln, G. A. 1987. Long-term stimulatory effects of a continuous infusion of LHRH agonist on testicular function in male red deer (*Cervus elaphus*). *J Reprod Fertil* 80:257–261.

Linn, G. S., and H. D. Steklis. 1990. The effects of depo-medroxyprogesterone acetate (DMPA) on copulation-related and agonistic behaviors in an island colony of stumptail macaques (*Macaca arctoides*). *Physiol Behav* 47:403–408.

Maclellan, L. J., E. G. M. Bergfield, L. A. Fitzpatrick, W. J. Aspden, J. E. Kinder, J. Walsh, T. E. Trigg, and M. J. D'Occhio. 1997. Influence of the luteinizing hormone-releasing hormone agonist, Deslorelin, on patterns of estradiol-17β and luteinizing hormone secretion, ovarian follicular responses to superstimulation with follicle-stimulating hormone and recovery and in vitro development of oocytes in heifer calves. *Biol Reprod* 56:878–884.

Mastrioanni, L. Jr., S. Suzuki, and F. Watson. 1967. Further observations on the influence of the intrauterine device on ovum and sperm distribution in the monkey. *Am J Obstet Gynecol* 99:649–660.

McShea, W. J., S. L. Monfort, S. Hakim, J. F. Kirkpatrick, I. K. M. Liu, J. W. Turner Jr., L. Chassy, and L. Munson. 1997. The effect of immunocontraception on the behavior and reproduction of white-tailed deer. *J Wildl Manag* 61:560–569.

Munson, L., J. E. Bauman, C. S. Asa, W. Jöchle, and T. E. Trigg. 2001. Efficacy of the GnRH-analogue deslorelin for suppression of the oestrous cycle in cats. *J Reprod Fertil Suppl* 57:269–273.

Nagle, C. A., and E. Turin. 1997. Contraception in bitches by non-surgical insertion of an intrauterine device (IUD). *Vet Argentina* 14:414–420.

Okkens, A. C., H. S. Kooistra, and R. F. Nickel. 1997. Comparison of long-term effects of ovariectomy versus ovariohysterectomy in bitches. *J Reprod Fertil Suppl* 51:227–231.

Orford, H. J. L. 1996. Hormonal contraception in free-ranging lions (*Panthera leo* L.) at the Etosha National Park. In *Contraception in Wildlife*, Book 1, ed. P. N. Cohn, E. D. Plotka, and U. S. Seal, 303–320. Lewiston, NY: Edwin Mellen.

Parborell, F., A. Pecci, O. Gonzalez, A. Vitale, and M. Tesone. 2002. Effects of a gonadotropin-releasing hormone agonist on rat ovarian follicle apoptosis: regulation by epidermal growth factor and the expression of Bcl-2-related genes. *Biol Reprod* 67:481–486.

Paterson, M., and R. Aitken. 1990. Development of vaccines targeting the zona pellucida. *Curr Opin Immunol* 2:723–747.

Penfold, L. M., R. Ball, I. Burden, W. Jöchle, S. B. Citino, S. L. Monfort, and N. Wielebnowski. 2002. Case studies in antelope aggression control using a GnRH agonist. *Zoo Biol* 21:435–448.

PDR (*Physicians' Desk Reference*). 2002. 56th edition. Montvale, NY: Medical Economics.

Pincus, G. 1965. *The Control of Fertility*. New York: Academic Press.

Pineda, M. H., and M. P. Dooley. 1984. Surgical and chemical vasectomy in the cat. *Am J Vet Res* 45:291–300.

Pineda, M. H., T. J. Reimers, L. C. Faulkner, M. C. Hopwood, and G. E. Seidel Jr. 1977. Azoospermia in dogs induced by injection of sclerosing agents into the caudae of the epididymides. *Am J Vet Res* 38:831–838.

Plotka, E. D., and U. S. Seal. 1989. Fertility control in deer. *J Wildl Dis* 25:643–646.

Plotka, E. D., T. C. Eagle, D. N. Vevea, A. L. Koller, D. B. Siniff, J. R. Tester, and U. S. Seal. 1988. Effects of hormone implants on estrus and ovulation in feral mares. *J Wildl Dis* 24:507–514.

Porton, I. 1995. Results for primates from the AZA contraception database: species, methods, efficacy and reversals. *Proceedings, Joint Conference of American Association of Zoo Veterinarians, Wildlife Disease Association, and American Association of Wildlife Veterinarians (AAZV/WDA/AAWV)*, 381–394, August 12–17, 1995. East Lansing, MI.

Portugal, M. M., and C. S. Asa. 1995. Effects of chronic melengestrol acetate contraceptive treatment on perineal tumescence, body weight, and sociosexual behavior of hamadryas baboons (*Papio hamadryas*). *Zoo Biol* 14:251–259.

Remfry, J. 1978. Control of feral cat populations by long term administration of megestrol acetate. *Vet Rec* 28:403–404.

Rijnberk, A., and J. A. Mol. 1997. Progestin-induced hypersecretion of growth hormone: an introductory review. *J Reprod Fertil Suppl* 51:335–338.

Seal, U. S., R. Barton, L. Mather, K. Oberding, E. D. Plotka, and C. W. Gray. 1976. Hormonal contraception in captive female lions (*Panthera leo*). *J Zoo Anim Med* 7:1–17.

Selman, P. J., J. A. Mol, G. R. Rutterman, E. van Garderen, T. S. G. A. N. van den Ingh, and A. Rijnberk. 1997. Effects of progestin administration on the hypothalamic-pituitary-adrenal axis and glucose homeostasis in dogs. *J Reprod Fertil Suppl* 51:345–354.

Shapiro, E. I., and S. J. Silber. 1979. Open-ended vasectomy, sperm granuloma and post-vasectomy orchalgia. *Fertil Steril* 32:546–550.

Sherwin, B. B., and M. M. Gelfand. 1989. A prospective one-year study of estrogen and progestin in postmenopausal women: effects on clinical symptoms and lipoprotein lipids. *Obstet Gynecol* 73:759–766.

Shirley, B., J. C. Bundren, and S. McKinney. 1995. Levonorgestrel as a post-coital contraceptive. *Contraception* 52:277–281.

Silber, S. J. 1979. Epididymal extravasation following vasectomy as a cause for failure of vasectomy reversal. *Fertil Steril* 31:309–315.

Silber, S. J. 1989a. Pregnancy after vasovasostomy for vasectomy reversal: a study of factors affecting long-term return of fertility in 282 patients followed for 10 years. *Hum Reprod* 4:318–322.

Silber, S. J. 1989b. Results of microsurgical vasoepididymostomy: role of epididymis in sperm maturation. *Hum Reprod* 4:298–303.

Silvestrini, B. 2002. Introduction to the special issue on control of male fertility and infertility. *Contraception* 65:257.

Simmons, J. G., and C. E. Hamner. 1973. Inhibition of estrus in the dog with testosterone implants. *Am J Vet Res* 34:1409–1419.

Sitruk-Ware, R. 2000. Progestins and cardiovascular risk markers. *Steroids* 65:651–658.

Sloan, J. M., and I. M. Oliver. 1975. Progestogen-induced diabetes in the dog. *Diabetes* 24:337–344.

Sokolowski, J. H., and S. Geng. 1977. Biological evaluation of mibolerone in the female beagle. *Am J Vet Res* 38:1371–1376.

Trigg, T. E., P. J. Wright, A. F. Armour, P. E. Williamson, A. Junaidi, G. B. Martin, A. G. Doyle, and J. Walsh. 2001. Use of a GnRH analogue implant to produce reversible long-term suppression of reproductive function in male and female domestic dogs. *J Reprod Fertil Suppl* 57:255–261.

Triman, K., and L. Liskin. 1988. Intrauterine device. *Popul Rep* 16:1–31.

Turin, E. M., C. A. Nagle, M. Lahoz, M. Torres, M. Turin, A. F. Mendizabal, and M. B. Escofet. 1997. Effects of a copper-bearing intrauterine device on the ovarian function, body weight gain and pregnancy rate of nulliparous heifers. *Theriogenology* 47:1327–1336.

VandeVoort, C. A., E. D. Schwoebel, and B. S. Dunbar. 1995. Immunization of monkeys with recombinant complementary deoxyribonucleic acid expressed zona pellucida proteins. *Fertil Steril* 64:838–847.

Vickery, B. H., G. I. McRae, J. C. Goodpasture, and L. M. Sanders. 1989. Use of potent LHRH analogues for chronic contraception and pregnancy termination in dogs. *J Reprod Fertil Suppl* 39:175–187.

Volpe, P., B. Izzo, M. Russo, and L. Iannetti. 2001. Intrauterine device for contraception in dogs. *Vet Rec* 149:77–79.

Webel, S. K., and E. L. Squires. 1982. Control of the oestrous cycle in mares with altrenogest. *J Reprod Fertil Suppl* 32:193–198.

Wildt, D. E., and D. F. Lawler. 1985. Laparoscopic sterilization of the bitch and queen by uterine horn occlusion. *Am J Vet Res* 46:864–869.

World Health Organization (WHO) Task Force for Epidemiological Research on Reproductive Health. 1994a. Progestogen-only contraceptives during lactation. I. Infant growth. *Contraception* 50:35–54.

World Health Organization (WHO) Task Force for Epidemiological Research on Reproductive Health. 1994b. Progestogen-only contraceptives during lactation. II. Infant development. *Contraception* 50:55–68.

Wright, P. J., J. P. Verstegen, K. Onclin, W. Jöchle, A. F. Armour, G. B. Martin, and T. E. Trigg. 2001. Suppression of the oestrous responses of bitches to GnRH analogue deslorelin by progestin. *J Reprod Fertil Suppl* 57:263–268.

Zimbelman, R. G., J. W. Lauderdale, J. H. Sokoloski, and T. G. Schalk. 1970. Safety and pharmacologic evaluations of melengestrol acetate in cattle and other animals: a review. *J Am Vet Med Assoc* 157:1528–1536.

CHERYL S. ASA

4
ASSESSING EFFICACY AND REVERSIBILITY

Assessing the effectiveness and reversibility of a contraceptive might seem a very straightforward process: there should be no young conceived and born during the period of contraceptive treatment, and young should be conceived and born following cessation of treatment. However, as is typical of biological systems, such assessments are actually rather complex, because a great number of factors bear on the likelihood of conception, pregnancy maintenance, and parturition, whether or not animals are or have been treated with contraceptives. This chapter reviews the most important of those factors, but does not cover all the variables subsumed under "individual differences," such as general health, immune response, metabolic rates, clearance rates, and individual levels of fertility.

Most basic to judging both efficacy and reversibility is an understanding of the points in the reproductive process at which the contraceptive acts, that is, whether it prevents pregnancy by suppressing follicle growth, or allows follicle growth but prevents ovulation, and so forth. A brief summary of the effects of the methods currently in use in most zoos is presented in Table 4.1 (for a more complete discussion of methods, see Chapter 3, Types of Contraception).

ASSESSING EFFICACY

Factors that must be considered in assessing whether a method is indeed effective include (1) whether the female was pregnant before the contraceptive was started, (2) latency to effect (how long after the method is initiated will it prevent conception or implantation), (3) whether the contraceptive is in still in place

Table 4.1
Contraceptive methods currently in use and their effects on reproductive processes

Compound or method	Products	Effects on reproductive processes
Synthetic progestins:		
Melengestrol acetate (MGA)	MGA implants, MGA feed	Prevent ovulation, thicken cervical mucus, interfere with sperm and egg transport, alter uterine endometrium
Medroxyprogesterone acetate	Depo-Provera injections	
Levonoregestrel	Norplant implants	
Megestrol acetate	Ovaban and Ovarid tablets	
Proligestone	Devosteron injections	
Altrenogest	Regu-mate oral solution	
Etonorgestrel	Implanon implants	
Combination estrogen–progestin:		
Ethinyl estradiol or megestrol plus levonorgestrel, desogestrel, noregestrel, norgestimate, or ethynodiol diacetate	Birth control pills (many formulations)	Effects of progestins alone, plus better suppression of follicle growth
Gonadotropin-releasing hormone (GnRH) agonists:		
Leuprolide acetate	Lupron Depot injection	After initial stimulation, suppress production and secretion of luteinizing hormone (LH) and follicle-stimulating hormone (FSH), which results in absence of stimulation of ovarian follicles or of testes; thus, there is no production of estradiol, progesterone, or testosterone, and thus no ovulation or spermatogenesis
Deslorelin	Suprelorin implants	
Porcine zona pellucida (PZP)	PZP vaccine	Prevents fertilization but allows follicle growth and ovulation
Intrauterine device (IUD)	ParaGard T 380A	Interferes with sperm and egg transport but allows follicle growth and ovulation

[implants, intrauterine device (IUD)], the injection was delivered, or the feed or pill was consumed, and (4) accuracy of the predicted duration of efficacy.

Importance of Pregnancy Detection

Although this is often overlooked, it is important to determine whether a female is pregnant at the time a contraceptive method is first administered, because conceptions that occur before initiation of treatment cannot be considered method failures. More critically, however, some methods may have deleterious effects on fetal development (for example, those containing an estrogen) or obstruct parturition (for example, those containing a progestin), whereas others may be abortifacient [IUDs and perhaps gonadotropin-releasing hormone (GnRH) agonists in some species, for example, Suprelorin: Peptech Animal Health; Lupron-Depot: TAP Pharmaceuticals]. The only method that currently seems safe for pregnant females is the porcine zona pellucida (PZP) vaccine, but it has not been evaluated in many species.

Latency to Effect

Contraceptives should not be considered effective immediately following initiation of oral administration, injection, or insertion of an implant. Because of the long history of steroid hormone use in humans, there are considerable data on the circulating levels of at least some of the progestins and progestin–estrogen combinations. Steroid hormone concentrations above that considered to be the threshold of efficacy may be reached within 24 hours (Howard et al. 1975; Alvarez et al. 1998). Depending on the stage of the cycle, however, that still may be too late to block the next ovulation. That is, once preovulatory processes are set in motion, a dose that might have prevented the initiation of those processes may still be insufficient to halt them (Barbosa et al. 1996; Brache et al. 1999; Hapangama et al. 2001). Thus, it is probably prudent either to keep the female separated from males or to use another form of contraception for the first week after the new treatment is begun (Petta et al. 1998).

The timeline for establishing effective titers of antibodies following inoculation with vaccines such as PZP is much more variable. Individual immune responses differ, and vaccine formulations also vary, with most requiring multiple doses or injections to achieve efficacy. In general, the doses should be spaced by as much as a month. Early formulations of PZP required as many as three injections, but currently two injections given 2 to 3 weeks apart seem sufficient (J. Kirkpatrick, personal communication). Still, more advance planning is required when using vaccines than steroid-based contraceptives.

An advantage to IUDs is that they can be considered effective immediately. Because IUDs function by interrupting transport of gametes and by impeding implantation of zygotes, even zygotes produced by fertile copulations that occur soon after IUD insertion can be prevented from implanting.

Castration and vasectomy (either surgical or chemical, such as with sclerosing agents) cannot be considered immediately effective, as sperm with the capacity to fertilize may remain in the tract (vas deferens) for weeks or even months (Pineda and Dooley 1984; Pineda et al. 1976; Cortes et al. 1997). Antispermatogenic agents such as bisdiamine, indenopyridene, or GnRH agonists may have an even longer latency to efficacy, because not only must the spermatogenic process be successfully halted, any sperm remaining in the vas must also be cleared. The process of sperm clearance can be hastened by inducing ejaculation (for example, electroejaculation), which also serves to document disappearance.

Ensuring Delivery

An important factor in assuring efficacy is certainty that the contraceptive was or is being reliably delivered. Implants deliver their contents continually, in some cases for years, but as with IUDs they can be expelled. Injectable vaccines or depot formulations similarly can be effective for long periods (for example, PZP vaccine, Depo-Provera, Lupron Depot), but a dart can fail to deliver a full dose. Animals can refuse to consume feed or swallow pills, and even surgical and chemical vasectomies have been known to reverse, or perhaps were never successful (Esho and Cass 1978; Cortes et al. 1997). Thus, steps should be taken to ensure and document delivery.

The possibility of implant loss can be reduced by using sterile insertion technique. Before insertion, melengestrol acetate (MGA) implants (available from E. D. Plotka) must be gas-sterilized (autoclaving destroys them and alcohol depletes the steroid) and then allowed to air, or degas, because residual gas may evoke a tissue reaction. Norplant (Wyeth) and Suprelorin implants are sterile as packaged, so do not require further preparation. When MGA implants that require a small incision for insertion are used, the animal may need to be separated from conspecifics during healing to prevent excessive grooming of the site.

Inserting implants where the animal is least likely to be able to self-groom, such as between the shoulder blades, also minimizes the chance of loss. MGA implants can even be sutured in place, a procedure not possible with other implants because it would interfere with their release dynamics. Identification transponder microchips inserted in MGA implants can be used to confirm presence and location, and stainless steel suture material may be incorporated into MGA implants to make them visible on a radiograph. Transponders and stainless steel cannot be added to other implants (for example, Norplant implants).

For contraceptives delivered by dart, recovering and inspecting the dart is usually adequate to confirm proper injection. However, darts can fire properly but still fail to inject their contents into the muscle. Working at sufficiently close range to document delivery may be necessary. When delivery appears to have failed or to have been incomplete, the procedure can be repeated. There are probably no deleterious effects associated with giving an additional dose of Lupron Depot or PZP, and this may also be true for Depo-Provera. However, high doses of progestins are probably best avoided because they also bind glucocorticoid receptors (Duncan et al. 1964; Fekete and Szeberenyi 1965; Selman et al. 1997) and are more likely to produce side effects at high doses, in particular, immunosuppression (Settepane et al. 1978).

Formulations delivered orally present another set of challenges, especially because they usually must be administered daily. For MGA incorporated in feed, the dosage must be carefully calculated to ensure that the animal will consume the entire amount, even when it is not particularly hungry. Thus, it is probably best that the MGA feed constitute only a portion of the total diet and that it be offered first. For herds, it may be necessary to either feed animals separately so that dominant animals do not monopolize the treated food or provide a higher concentration to increase the probability that all individuals ingest an adequate dose.

Contraceptives in pill form often must be incorporated into a treat in some way to disguise its presence and ensure ingestion. Primates in particular may hold a pill in the cheek, only to spit it out later. Crushing the pill in syrup, liquid, or other favorite food may avoid this problem.

In humans, IUDs that are properly inserted by an experienced physician are seldom expelled (Chi 1993). However, because nonhuman mammalian uteri vary considerably, expulsion may be more common. In particular, the long tails provided on some IUDs should be cut to prevent IUD removal by the animal.

Monitoring the reproductive processes affected by the contraceptive is another approach to assessing whether the product was successfully delivered or is still in place. For example, for methods that suppress follicle growth, the subsequent absence of estradiol will be evidenced by an absence of estrous behavior or other signs of estrus, such as perineal swelling or reddening in some primates, vaginal opening in some prosimians and rodents, vulval swelling and vaginal discharge in many species, and changes in vaginal cytology (see Asa 1996 for review). In many female primates and bats, the absence of menstruation or the sanguinous discharge of some proestrous and estrous canids (Asa 1998) is indicative of ovarian suppression.

Duration of Efficacy

As already mentioned, orally active progestins must be administered daily to be effective, and missing a dose for 1 or 2 days may well result in conception. This rapid restoration of fertility is probably caused by the relative ineffectiveness of progestins in suppressing follicular growth (Broome et al. 1995; Alvarez et al. 1996). Once the progestin is removed, the follicles can quickly enter the final growth phase and ovulate. In fact, such a response is the basis for estrous synchronization protocols in many domestic species (Adams et al. 1992). In contrast, because the estrogen component of the combination birth control pills is more effective at suppressing follicle growth, even the weeklong placebo or pill-free period is not sufficient for completion of the follicular growth phase and ovulation (Mall-Haefeli et al. 1988). However, missing combination pills during the other times of the month may allow enough follicular growth to risk possible ovulation during the subsequent hormone-free week. Thus, any oral contraceptives should be administered carefully according to directions or conception may occur.

The duration of efficacy of injectable depot preparations such as Depo-Provera and Lupron Depot and for implants such as MGA, Norplant, and Suprelorin is quite variable. The vehicle or matrices that release the active ingredients seem to have different dynamics in different individuals, making it difficult to estimate accurately the length of time in which the contraceptive compound is circulating at levels above the threshold of efficacy. Although the mechanism of action of vaccines differs considerably from that of the depot preparations or implants, the duration of efficacy of vaccines is also notoriously variable because of individual differences in immune response. For these products, length of effect must be expressed as a range, based on reports from treated animals. For this reason, when estimating the period of contraception, the minimum duration of efficacy is typically given. However, when planning for reversal, the maximum duration of efficacy becomes more important.

An additional variable not often considered when calculating dose and dose intervals for implants and injectable hormones is that the frequency required for administering successive doses is a function not only of the vehicle or matrix but also of the amount of hormone or dose. To be effective, the circulating concentration of the hormone must be above the threshold of efficacy. Following delivery of a higher dose, the circulating concentration will remain above that threshold for a longer period of time. Thus, if a female shows signs of estrus only 6 weeks after a Depo-Provera injection when 2 or 3 months of suppression were expected, either the dose could be increased or the interval shortened to achieve the same effect.

A misunderstanding about the period of efficacy of implants can lead to erroneous conclusions regarding their reversibility. The recommendation for timing implant replacement is always set conservatively; that is, the minimum duration of efficacy that has been observed for the method in a particular species is used to establish the schedule for implant replacement to ensure efficacy. However, some people have interpreted that recommendation to mean that the implants are depleted and no longer effective in any individual after the stated interval. This assumption has led to the mistaken conclusion that the method may not be reversible (De Vleeschouwer et al. 2000; DeMatteo et al. 2002).

ASSESSING REVERSIBILITY

Measuring time to reversal varies for many reasons. First, there are different points in the reproductive process that can be used as the measure of successful reversal. The most definitive, of course, is the birth of a healthy infant. However, this is not as simple an assumption as it might seem because there are many steps in the process that can fail independent of contraceptive history. Even females who have never been contracepted, when placed with a male, may not become pregnant during years of cohabitation. Thus, pregnancy rates postcontraception must always be compared not only to precontraceptive reproductive history but also to the pregnancy rates of noncontracepted females that are matched at least for age and parity (number of previous young). Also, it must not be forgotten that failure to produce young may be attributable to male infertility or genetic incompatibility within the pair . . . it does indeed take two!

Many factors other than contraceptives affect the likelihood of ovulation and conception. These variables include reproductive history, age, health, body weight (very thin or obese animals may not ovulate or conceive), season, social status, and, of course, fertility of the partner. Although it should be obvious, access to a behaviorally compatible, fertile partner during the period being evaluated is the most basic criterion. The countless intrinsic and environmental factors that influence the probability of successful reproduction argue against relying on production of live young as the sole measure of reversibility.

Latency to Clearance of the Contraceptive

For many kinds of contraception, the most basic measure of reversal is the decrease in concentration of the contraceptive compound, or titer in the case of vaccines, in the body below the threshold of efficacy. The circulating hormones from implants (MGA; Norplant), oral preparations containing a progestin (MGA, Ovaban; Schering Plough), and birth control pills clear very rapidly following removal or cessation of administration (Kook et al. 2002; Buckshee et al. 1995), so that ovulation and conception may occur within days, although actual latency is usually longer and depends on the individual.

Latency to clearance following depot formulations delivered by injection (Depo-Provera, Lupron Depot) can vary greatly among individuals. For example, time to first ovulation following the last injection of Depo-Provera ranges from 6 weeks to 2 years (Schwallie and Assenzo 1974; Nash 1975; Ortiz et al. 1977). Similarly, because Suprelorin implants degrade slowly and cannot be removed, reversal cannot be predicted except in very general terms (typically, plus or minus several months). The situation is comparable with vaccines such as PZP, because individual immune responses vary considerably, and in fact contraception may not be reversible in some species or in those treated for more than 3 years (see Chapter 3, Types of Contraception).

Latency to Resumption of Ovarian Cycles

Because it is often not possible to measure concentrations of the contraceptive drug, confirmation that reversibility has been successful often must depend on documentation of other evidence that reproductive potential has been restored. For methods that inhibit ovulation, verification that ovulatory cycles have resumed may be an adequate measure. In cases in which gonadal hormones can be assayed, sustained elevations in progesterone for the length of the species-specific luteal phase indicate ovulation. Alternatively, ultrasound evidence of follicle growth and rupture followed by corpus luteum formation provides more direct confirmation of ovulation. Short of hormone assay or ultrasound exams, signs of estrus (indirect indicators of circulating estradiol) may be sufficient to document return to ovarian activity. These signs include, depending on the species (see earlier section on Ensuring Delivery), estrous behavior, perineal swelling or reddening, vulval swelling, vaginal opening or discharge, changes in vaginal cytology, and proestrous bleeding (some canids). In addition, resumption of regular menstruation signals ovulation in some primates and bats.

Species with induced ovulation present a special case for documenting complete resumption of ovarian activity. Those females not housed with males will not experience the copulatory activity necessary for ovulation, so measurement of estradiol or regular waves of follicle growth would be sufficient to indicate resumption of ovarian activity following depletion of a deslorelin implant. However, for progestin-based contraceptives that do not typically completely suppress such waves of follicular growth in either induced or spontaneous ovulators, these measures are not adequate for assessing either efficacy or reversal. Instead, an increase in progesterone following copulation or ultrasound evidence of follicle rupture and corpus luteum formation is needed.

Latency to First Conception

Including a category for conception that may not result in a live birth may appear being pessimistic about the consequences of contraception. However, for species in which early pregnancy diagnosis is possible (many laboratory and domestic species, as well as humans), the data suggest that, even in females never contracepted, between 20 and 66 percent of embryos are lost spontaneously, most often within the first weeks following conception (Perry 1954; Smart et al. 1982; Wilmut et al. 1986; McRae 1992). The reasons for such high rates of loss have not all been documented, but genetic abnormalities are suspected in many cases. Thus, the true latency to first conception following cessation or removal of contraceptive treatment may well be a shorter period than when the criterion for success is the birth of young.

Latency to Birth of Young

Despite the numerous factors that affect the likelihood of pregnancy and parturition, the most conservative measure of successful contraceptive reversal is the birth of live young. However, the complexity of such a measure was illustrated in an elegant study that evaluated reversibility of MGA implants in golden lion tamarins (Wood et al. 2001). Variables considered included previous reproductive history, age when placed in a breeding situation, date of reproduction (that is, birth of young), the date and reason data collection ended (which included death of the female or her mate), removal from the breeding situation, reimplantation with MGA, reproductive senescence, and end of the study. Another critical factor was establishing the start time (time 0) for calculating latency to reversal, which was designated as the time when the female was placed in a breeding situation. Survival analysis was used because it accounted for the amount of time each female was in the study, as individual females entered (had implants removed and were placed in a breeding situation) and left (death, reimplantation, etc.) the study at different times.

After MGA implant removal, 75 percent of the female golden lion tamarins reproduced within 2 years (Wood et al. 2001), a rate that was indistinguishable from that of control females that had never been implanted with MGA. However, the reproductive rate was, not surprisingly, much lower in females with implants left in place beyond 2 years, demonstrating that the actual duration of MGA efficacy was more than 2 years.

Effect of Contraceptives on Litter Size and Infant Survival

When calculating time to first reproduction following cessation or removal of contraception, that time is typically assumed to be the date the female gives birth. However, this measure is confounded by the possibility of stillbirth. Because there are so many factors independent of prior contraception that affect the development, survival, and delivery of a fetus, simply giving birth might still be considered successful reproduction. Although such an assumption may appear reasonable, the study by De Vleeschouwer et al. (2000) found that stillbirths were more common in female golden-headed lion tamarins previously implanted with MGA. However, they did not separate females with implants removed from those with implants still in place and presumed to have expired (that is, in place more than 2 years).

Wood et al. (2001) did not find an association between prior MGA implant use and stillbirths or litter size in female golden lion tamarins when they separated those with implants removed from those with implants in place more than 2 years. However, they did detect a decrease in infant survival in seven females following MGA implant removal compared to survival rate of their infants before MGA use. Although this result raises questions about possible effects of MGA use on subsequent reproductive success, the authors noted that these seven females had higher infant survival rates than the control females before MGA treatment but that their rates were not different from controls following MGA use. In another callitrichid species, the common marmoset (*Callithrix jacchus*), MGA contraception was associated with hypertrophied and decidualized uteri, but this condition reversed when the implants were removed (Möhle et al. 1999). All females ovulated within 3 weeks and conceived within 4 months. All but one gave birth to live young with normal litter size and sex ratio; one female aborted because of a vaginal infection, which is very unlikely to be attributable to past contraceptive use.

Because observed numbers were so small, it is important to continue to monitor reproductive rates following MGA use in these as well as other species. As pointed out in the chapters in Part III (this volume) that analyze the Contraception Database, there have been so few attempts at reversal per species that adequate analyses have been impossible in all but a few cases.

Latency to Appearance of Sperm in the Ejaculate

For methods that suppress spermatogenesis, verification of the reappearance of sperm in the ejaculate that is of the quality and concentration typical of the species demonstrates reversal. However, the first step in this process is resumption of spermatogenesis, which will be affected by such factors as age, social status, and time of year, that is, whether the contraceptive is withdrawn during the breeding season. For methods such as deslorelin implants, there is the additional uncertainty of duration of efficacy, because the implants gradually degrade as the hormone is released. Thus, it is not possible to accurately predict when the concentration of the contraceptive will fall below the threshold of efficacy unless blood levels of deslorelin are assayed.

Following resumption of spermatogenesis, there is a species-specific time for completion of the process before mature sperm finally are released into the vas deferens. For most mammals, the time to sperm production ranges from 6 to 8 weeks. Thus, after the contraceptive is removed or is no longer effective, it may be as long as 2 months before sperm can be expected in the ejaculate.

REFERENCES

Adams, G. P., R. L. Matteri, and O. J. Ginther. 1992. Effect of progesterone on ovarian follicles, emergence of follicular waves and circulating follicle-stimulating hormone in heifers. *J Reprod Fertil* 96:627–640.

Alvarez, F., V. Brache, A. Faundes, A. S. Tejada, and F. Thevenin. 1996. Ultrasonographic and endocrine evaluation of ovarian function among Norplant implant users with regular menses. *Contraception* 54:275–279.

Alvarez, F., V. Brache, A. S. Tejada, L. Cochon, and A. Faundes. 1998. Sex hormone binding globulin and free levonorgestrel index in the first week after insertion of Norplant® implants. *Contraception* 58:211–214.

Asa, C. S. 1996. Reproductive physiology. In *Wild Mammals in Captivity*, ed. D. G. Kleiman, M. E. Allen, K. V. Thompson, and S. Lumpkin, 390–417. Chicago: University of Chicago Press.

Asa, C. S. 1998. Dogs (Canidae). In *Encyclopedia of Reproduction*, ed. E. Knobil and J. D. Neill, 80–87. New York: Academic Press.

Barbosa, I. C., E. Coutinho, C. Hirsch, O. A. Ladipo, S. E. Olsson, and U. Ulmsten. 1996. Temporal relationship between Uniplant insertion and changes in cervical mucus. *Contraception* 5:213–217.

Brache, V., P. D. Blumenthal, F. Alvarez, T. R. Dunson, L. Cochon, and A. Faundes. 1999. Timing of onset of contraceptive effectiveness in Norplant® implant users. II. Effect on the ovarian function in the first cycle of use. *Contraception* 59:245–251.

Broome, M., J. Clayton, and K. Fotherby. 1995. Enlarged follicles in women using oral contraceptives. *Contraception* 52:13–16.

Buckshee, K., P. Chatterjee, G. I. Dholl, M. N. Hazra, B. S. Kodkany, K. Lalitha, A. Logambal, P. Manchande, U. K. Nande, G. RaiChoudhury, P. Rajarem, P. C. Sengupta, R. Sivaramen, U. D. Sutaria, K. Zaveri, S. Datey, L. N. Gaur, N. K. Gupta, S. Mehta, M. Roy, N. C. Saxene, K. Topo, S. C. Yadav, and P. Vishwaneth. 1995. Return to fertility following discontinuation of Norplant®. II. Subdermal implants. *Contraception* 51:237–242.

Chi, I. 1993. What we have learned from recent IUD studies: a researcher's perspective. *Contraception* 48:81–108.

Cortes, M., A. Flick, M. A. Barone, R. Amatya, A. E. Pollack, J. Otero-Flores, C. Juarez, and S. McMullen. 1997. Results of a pilot study of the time to azoospermia after vasectomy in Mexico City. *Contraception* 56:215–222.

DeMatteo, K. E., I. J. Porton, and C. S. Asa. 2002. Comments from the AZA Contraception Advisory Group on evaluating the suitability of contraceptive methods in golden-headed lion tamarins (*Leontopithecus chrysomelas*). *Anim Welf* 11:343–348.

De Vleeschouwer, K., K. Leus, and L. Can Elsacker. 2000. An evaluation of the suitability of contraceptive methods in golden-headed lion tamarins (*Leontopithecus chrysomelas*), with emphasis on melengestrol acetate (MGA) implants. (I) Effectiveness, reversibility and medical side-effects. *Anim Welf* 9:251–271.

Duncan, G. L., S. C. Lyster, J. W. Hendrix, J. J. Clark, and H. D. Webster. 1964. Biologic effects of melengestrol acetate. *Fertil Steril* 15:419–432.

Esho, J. O., and A. S. Cass. 1978. Recanalization rate following methods of vasectomy using interposition of fascial sheath of vas deferens. *J Urol* 120:178–179.

Fekete, G., and S. Szeberenyi. 1965. Data on the mechanism of adrenal suppression by medroxyprogesterone acetate. *Steroids* 6:159–166.

Hapangama, D., A. F. Glasier, and D. T. Baird. 2001. The effects of peri-ovulatory administration of levonorgestrel on the menstrual cycle. *Contraception* 63:123–129.

Howard, G., R. J. Warren, and K. Fotherby. 1975. Plasma levels of norethistrone in women receiving norethistrone oenanthate intramuscularly. *Contraception* 12:45–50.

Kook, K., H. Gabelnick, and G. Duncan. 2002. Pharmacokinetics of levonorgestrel 0.75 mg tablets. *Contraception* 66:73–76.

Mall-Haefeli, M., I. Werner-Zodrow, P. R. Huber, and A. Edelman. 1988. Oral contraception and ovarian function. In *Female Contraception*, ed. B. Runnebaum, T. Rabe, and L. Kiesel, 97–105. Berlin: Springer-Verlag.

McRae, A. C. 1992. Observation on the timing of embryo mortality in ranch mink (*Mustela vison*). *Proc 40th Annu Meet Canada Mink Breeders Assoc* 1992:3–48.

Möhle, U., M. Heistermann, A. Einspanier, and J. K. Hodges. 1999. Efficacy and effects of short- and medium-term contraception in the common marmoset (*Callithrix jacchus*) using melengestrol acetate. *J Med Primatol* 28:36–47.

Nash, H. A. 1975. Depo-Provera: a review. *Contraception* 12:377–393.

Ortiz, A., M. Hiroi, F. Z. Stanczyk, U. Goebelsmann, and D. R. Mishell Jr. 1977. Serum medroxypreogesterone acetate (MPA) concentration and ovarian function following intramuscular injection of Depo-MPA. *J Clin Endocrinol Metab* 44:32–38.

Perry, J. S. 1954. Fecundity and embryonic mortality in pigs. *J Embryol Exp Morphol* 2:308–322.

Petta, C. A., A. Faundes, T. R. Dunson, M. Ramos, M. DeLucio, D. Faundes, and L. Bahamondes. 1998. Timing of onset of contraceptive effectiveness in Depo-Provera users. II. Effects on ovarian function. *Fertil Steril* 70:817–820.

Pineda, M. H., and M. P. Dooley. 1984. Surgical and chemical vasectomy in the cat. *Am J Vet Res* 45:291–300.

Pineda, M. H., T. J. Reimers, and L. C. Faulkner. 1976. Disappearance of spermatozoa from the ejaculates of vasectomized dogs. *J Am Vet Med Assoc* 168:502–503.

Schwallie, P. C., and J. R. Assenzo. 1974. The effect of depo-medroxyprogesterone acetate on pituitary and ovarian function, and return of fertility following its discontinuation: a review. *Contraception* 10:181–197.

Selman, P. J., J. A. Mol, G. R. Rutterman, E. van Garderen, T. S. G. A. M. van den Ingh, and A. Rijnberk. 1997. Effects of progestin administration of the hypothalamic-pituitary-adrenal axis and glucose homeostasis in dogs. *J Reprod Fertil Suppl* 51: 345–354.

Settepane, G. A., R. K. Pudupakkam, and J. H. McGowen. 1978. Corticosteroid effect on immunoglobulins. *J Allergy Clin Immunol* 62:162–166.

Smart, Y. C., L. S. Fraser, T. K. Roberts, R. L. Clancy, and A. W. Cripps. 1982. Fertilization and early pregnancy loss in healthy women attempting conception. *Clin Reprod Fertil* 1:177–184.

Wilmut, I., D. I. Sales, and C. J. Ashworth. 1986. Maternal and embryonic factors associated with prenatal losses in mammals. *J Reprod Fertil* 76:851–864.

Wood, C. W., J. D. Ballou, and C. S. Houle. 2001. Restoration of reproductive potential following expiration or removal of melengestrol acetate contraceptive implants in golden lion tamarins (*Leontopithecus rosalia*). *J Zoo Wildl Med* 32:417–425.

LINDA MUNSON, ANNEKE MORESCO, AND PAUL P. CALLE

5
ADVERSE EFFECTS OF CONTRACEPTIVES

Contraception is necessary to manage reproduction in zoo animals and to limit reproduction in free-ranging wild populations whose reproductive potential exceeds available space and resources. Although efficacy is important, contraceptive safety becomes a priority when contraceptives are used in high-profile zoo animals or for threatened or endangered species, when use in wildlife is controversial, or when the risk of adverse effects of the contraceptive is significant. As no medical procedure or drug treatment is completely without risk, a zero-tolerance attitude regarding adverse effects is not realistic. However, the advantages of contraception must outweigh the disadvantages to be acceptable.

Most commercially available contraceptives were designed to regulate human reproduction and have been adequately tested for safety only in humans and in laboratory animals. Federal Food and Drug Administration (FDA) regulations require clinical testing in two mammalian species, one of which must be a nonrodent species, before a compound can proceed to human clinical trials (Food and Drug Administration 1997). Consequently, most contraceptives have been tested for side effects in rats, mice, primates, or dogs. Safety for captive and free-ranging wildlife has been inferred from these laboratory animal studies, and in some cases this extrapolation may be inappropriate. Reproductive structure and function are different in each species, particularly in females, as factors controlling ovarian cyclicity, endometrial growth, and endometrial function vary temporally and quantitatively among species. However, testing contraceptive safety in the large number of species for which there is a current need for contraception is not practical. Extra-label use of contraceptives in wild animals has been necessary because of inadequate funding for clinical trials and the difficulty of obtaining sufficient numbers of appropriate animals within a species for safety assessments. Factors such as differences in age, parity, diet, and concurrent diseases confound interpretation of results, and the comprehensive pathological analyses required by the FDA cannot be conducted without sacrificing animals, which is not acceptable in rare and endangered species. Thus, all past and current contraceptive use in captive and free-ranging wild animals can be viewed as part of an ongoing global clinical trial that we are now monitoring to assess health risks for captive and free-ranging wildlife.

Reproduction depends on exquisitely timed, intricately connected, dose-dependent endocrine events that are species specific (Clark and Mani 1994; Hafez et al. 2000). Gonadal hormones such as estrogen, progesterone, and testosterone not only orchestrate activities throughout the reproductive tract but also have physiological effects on other tissues. Fertilization requires an intact reproductive tract and specific physical interactions between sperm and egg. Contraceptives are designed to disrupt reproduction by interfering with endocrine cyclicity or feedback, gamete production, or fertilization. In a broader context, compounds or devices that prevent implantation are also considered contraceptives. Endocrine cyclicity can be disrupted by adding hormones, preventing their synthesis or release, or interfering with hormone activity at the tissue level. Gamete production can be prevented by altering hormone levels or by targeting sperm development with chemicals. Fertilization can be averted by physically blocking the reproductive tract or by preventing the interaction of sperm and egg with antibodies. Designing contraceptives that disrupt only reproductive function, yet are safe, reversible, and effective in many species, has been the challenge. Any compound that disrupts endocrine function has the potential to disturb metabolic homeostasis and thereby cause disease. Antibodies that target sperm or egg proteins have the potential to incite immune-mediated damage in organs that express similar proteins or in surrounding tissues (Bagavant et al. 2002; Dunbar et al. 1989). These potential and actual risks need to be assessed and then weighed against the benefits of contraceptives in a specific animal or population. Side effects that are acceptable in free-ranging animals may not be tolerated in captive animals, and vice versa. For example, a serious disease resulting from contraception in an individual zoo animal would likely not be acceptable, whereas the health of an individual animal might be a less critical concern in wildlife. Conversely, a contraceptive that altered behavior or had a long-term effect on population structure would be inappropriate for free-ranging populations, but the effects might be manageable in captive animals and therefore be of lesser concern for that application.

The Contraceptive Health Surveillance Center for Animals was initiated in 1985 to meet the need for centralizing information on adverse effects of contraceptives in animals and became part of the American Zoo and Aquarium Association (AZA) Contraception Advisory Group (CAG) in 1990. The Center capitalized on the worldwide contraceptive "experiment in progress" in wildlife by requesting reproductive tracts from all captive and free-ranging wildlife to determine the pathological effects of contraceptives on reproductive health. The Center surveys reproductive tracts from both animals that have been treated with contraceptives and those which have not, so that the effect of a treatment can be recognized against the background of spontaneous reproductive diseases in a species. In the years since its inception, reproductive tracts from more than 1,500 mammals, including felids, canids, ursids, and other carnivores, primates, ungulates, and marine mammals, have been examined. The Center also maintains a database of published accounts of adverse health effects of contraceptives and serves as a reporting site for suspected side effects. Furthermore, the Center has conducted several clinical trials to evaluate contraceptive safety, including porcine zona pellucida (PZP) vaccination trials in felids and cervids, bisdiamine (an antispermatogenic compound) in canids and felids, and the gonadotropin-releasing hormone (GnRH) agonist deslorelin (Suprelorin; Peptech Animal Health) in felids. This chapter reviews information derived from the Center's surveillance and clinical studies as well as published accounts of adverse reactions from other investigators. For this chapter, contraception implies reversibility, and methods that permanently prevent reproduction (that is, sterilization) are not covered. It should be noted that there are many other published contraceptive studies that are not included in this review because safety was not specifically evaluated. The authors also refer readers to an excellent review on potential problems with wildlife contraception (Nettles 1997). Further, the AZA Wildlife Contraception Center website (www.stlzoo.org/contraception) has information on adverse effects arranged by taxon.

PROGESTIN CONTRACEPTIVES

Progestin contraceptives include melengestrol acetate (MGA), megestrol acetate (Megace, Ovaban: Schering-Plough), levonorgestrel (Norplant: Wyeth), and medroxyprogesterone acetate (MPA) (Provera and Depo-Provera: Pharmacia and Upjohn). MGA and levonorgestrel, the contraceptives most widely used in captive animals (see Appendix), have also occasionally been used in free-ranging animals (Orford and Perrin 1988; White et al. 1994). MGA, delivered in a Silastic subcutaneous implant or in feed, provides continual exposure to a potent progestin (Lauderdale 1983) that can have profound effects on reproductive health and metabolic function in some species. Most information on progestin side effects has been derived from MGA-treated animals.

General Effects of Progestins on the Reproductive Tract

Progesterone, the hormone of pregnancy, maintains the female reproductive tract in a state of nurturing (Clark and Mani 1994; Stock and Metcalfe 1994; Hafez et al. 2000). Briefly, the endometrial lining of the uterus grows in synchrony with the fetus and secretes nutritive fluids. In some species, the typically fibrous endometrial stromal cells transform into plump endocrine-appearing cells (decidual response). The local immune function is abated to prevent rejection of the foreign graft (the fetus) (Hansen 2000). The cervix closes, and the myometrium becomes quiescent and loses tone. Progesterone promotes maternal anabolic metabolism and also feeds back on the hypothalamus and pituitary gland, suppressing GnRH, follicle-stimulating hormone (FSH), and luteinizing hormone (LH) secretion. Additionally, progesterone prepares the mammary gland for lactation by promoting development of new glandular tissue (ductal and alveolar) and stimulating secretion of milk.

Synthetic progestins used for contraceptives have effects on the reproductive tract similar to endogenous progesterone, the magnitude of which depends on their relative potency. These physiological activities of progestins are the basis for the lesions that occur in progestin-contracepted animals. Although endometrial growth and fluid production are beneficial during pregnancy, they can become pathological over time in the absence of pregnancy. A uterus without tone and with suppressed immune function, as a result of progestins, is predisposed to develop endometritis (uterine infection) and pyometra (accumulation of pus in the uterus). Progestin-induced mammary growth can lead to tumor formation when lactation does not ensue (Mol et al. 1996). Progestins also can worsen subclinical diabetes by inhibiting insulin secretion (Straub et al. 2001). The effect of progestin-induced uterine lesions usually is only infertility. However, mammary and uterine cancer, pyometra, or diabetes can be fatal. Because the character and severity of lesions that occur with progestins are similar among species within a taxonomic group, the effects of progestins are summarized by taxonomic order in the following sections.

Carnivora

Carnivores appear exquisitely sensitive to progestin-induced diseases. In a study of zoo felids, progestins significantly increased the risk of developing moderate to severe endometrial hyperplasia (excessive growth of the uterine lining), endometrial mineralization (deposition of mineral in the uterine lining), and hydrometra (accumulation of fluid in the uterus), conditions that would cause permanent infertility in affected animals even after removal of the implant (Munson et al. 2002b). The risk of acquiring these reproductive diseases was significantly greater in felids treated with MGA for more than 72 months. Furthermore, felids treated with MGA also developed endometrial hyperplasia earlier in life. Other proliferative and inflammatory lesions of the endometrium, such as endometrial polyps, endometritis, or pyometra, were also more common in MGA-treated felids than in untreated animals, but the greater risk of developing these conditions was associated with endometrial hyperplasia and not with MGA exposure alone (Munson et al. 2002b). Regardless, these secondary conditions would further contribute to infertility and possibly poor health. Prevalence of endometrial cancer is also unusually high in zoo felids on progestins, with 27 of 28 cases submitted to the Center and another case report (Linnehan and Edwards 1991) having occurred in MGA-treated animals.

Although leiomyomas (benign smooth muscle tumors) were common in the population of zoo felids submitted to the Center, MGA did not increase the risk of developing these tumors (Chassy et al. 2002). However, five felids in this study had fatal leiomyosarcomas (malignant smooth muscle tumors), and four of these animals were treated with MGA contraceptives. Whether MGA promotes malignancy in smooth muscle tumors should be investigated. MGA exposure also increased the risk of mammary cancer in zoo felids, and these cancers had a rapid, aggressive clinical course resulting in death (Harrenstien et al. 1996).

MGA treatment did not prevent folliculogenesis or ovulation in zoo felids at doses used in implants (Kazensky et al. 1998). This failure to suppress ovarian function is a cause for concern because exposure to endogenous estrogens and progesterone can also cause proliferative lesions in the endometrium and mammary glands, and these hormones could further intensify the effects of MGA or other exogenous progestins. Primary ovarian tumors were not more common in MGA-treated felids, although metastases to the ovary from mammary cancer were more prevalent in MGA-treated animals, as would be expected because these cancers were associated with progestin exposure. MGA-treated felids also tended to have more cystic rete ovarii (nonhormonally active cysts within the ovary that destroy ovarian tissue as they expand), although this association was not statistically significant for the sample size in this study (Kazensky et al. 1998). Because progestins promote secretion of the epithelium lining these cysts, long-term progestin exposure would be expected to worsen this condition. Compression damage to the ovary caused by cystic rete ovarii would result in infertility.

Cystic endometrial hyperplasia and pyometra occur commonly in progestin-treated zoo canids (L. Munson, unpublished data), although the risk to animals treated with contraceptives of developing disease, above that which occurs spontaneously in canids, has not yet been measured. Mammary carcinomas also have been observed in MGA-treated canids; however, the increased risk from MGA treatment relative to nulliparity has not been assessed for wild canids. Endometrial hyperplasia occurs in MGA-treated ursids, but hyperplasia also occurs spontaneously in this taxon. The sample size of ursid cases submitted to the Center is too small at this time to accurately assess whether the risk of endometrial hyperplasia is greater in progestin-treated bears. Endometrial hyperplasia, pyometra, and endometrial cancers have been noted in other MGA-treated carnivores, but at this time tissue from too few animals has been submitted to the Center to determine whether these lesions are spontaneous in these species or caused by MGA.

Anecdotal reports of weight gain in carnivores treated with progestins have been submitted to the Center; whether this is a direct effect of progestins or secondary to decreased activity from estrus suppression remains to be evaluated. Several felids on progestins have developed diabetes, and a causal relationship to progestin treatment was suggested by improvement after removal of the implant in one animal (Kollias et al. 1984). Androgenization has also been reported in a few progestin-treated female lions, manifested by the development of a mane (Seal et al. 1976).

Ungulates

Endometrial hyperplasia, hydrometra, and uterine infections (endometritis and pyometra) have occurred in MGA-treated ungulates. However, these uterine diseases also occur spontaneously in this taxon, and it is not known whether progestins increase the risk. White-tailed deer (*Odocoileus virginianus*) treated with progestins had a prolonged breeding season, which was suspected to have caused weight loss in the herd, increased trauma to males, and birth of fawns in late summer and fall, timing that compromised their viability (Turner et al. 1992). In another study, progestin-treated deer grazed less frequently and lost weight (White et al. 1994). As MGA is now available in feed, a larger data set will be available in the future for analysis of the role of progestins in the development of diseases and behavioral changes in ungulates. One study reported some antler malformations and delayed or aberrant shedding of velvet development in male barasingha (*Cervus duvauceli*) and sambar (*Cervus unicolor*), but no other adverse effects were noted (Raphael et al. 2003).

Primates

Primates are a diverse group, ranging from Lemuridae with a bicornuate uterus and epitheliochorial placentation to great apes with a simple uterus and hemochorial placentation (Mossman 1987). Because the response of the endometrium to progestins depends on whether a species has epitheliochorial, endotheliochorial, or hemochorial placentation, progestin contraceptives would be expected to affect each primate group differently. Progestin-only and estrogen–progestin combination contraceptives have been extensively safety tested (for human use) in macaques, and widespread clinical application in humans has provided abundant data on contraceptive safety. Because great ape reproductive function is similar to that of humans (Mossman 1987), these safety studies should be applicable to apes. In support of this assumption, combination estrogen–progestin oral contraceptives have been used in apes without any adverse effects reported to date. Conversely, extrapolation of human safety assessments to primates other than macaques or apes may not be appropriate. The Center is in the process of accruing pathology information on all primates to evaluate the risk of long-term progestins in primate genera.

Some Lemuridae and Callitrichidae develop endometrial hyperplasia spontaneously as they age, but appear to contract more significant disease when treated with MGA contraceptives (L. LaFranco, personal communication; L. Munson, unpublished data). Whether endometrial hyperplasia impairs fertility in these species is not known. Most apes and Old World monkeys appear resistant to these proliferative effects, but develop endometrial atrophy (quiescence) when on long-term progestin therapy (L. Munson, unpublished data). This atrophy is likely to be reversible and would not affect fertility. Goeldi's monkeys and squirrel monkeys respond differently to MGA contraceptives by developing an exuberant, aggressive decidual response and secondary endometritis (Murnane et al. 1996). Therefore, MGA contraceptives are not recommended for Goeldi's monkeys and should be used with caution in squirrel monkeys.

Cases of diabetes mellitus in black macaques (*Macaca nigra*) and mangabeys (*Cercocebus* sp.) receiving MGA contraceptives have been reported to the Center. Whether diabetes is associated with progestins in these species is not known, because they are predisposed to develop diabetes from islet amyloidosis (O'Brien et al. 1993, 1996). In humans, progestins can exacerbate subclinical diabetes through inhibition of insulin secretion (Straub et al. 2001), which may also be the mechanism in macaques. The potential for diabetes to develop in progestin-treated primates needs further study so managers can weigh these risks against the benefit of contraception. Metabolic changes manifested by darkening of the pelage and weight gain have been noted in the black lemur (*Eulemur macaco*) on MPA (Porton 1995). Weight gain was also reported in progestin-treated hamadryas baboons (*Papio hamadryas*; Portugal and Asa 1995).

Pinnipeds and Cetaceans

Progestins have been used for contraception in marine mammals, but currently data are insufficient to determine if these exogenous hormones have adverse effects on reproductive or general health.

GONADOTROPIN-RELEASING HORMONE AGONISTS

Estrogens and progestins, whether endogenous or exogenous, incite pathological growth in the reproductive tissues of many species over time. Therefore, an ideal contraceptive would suppress ovarian function temporarily, providing reproductive rest to the gonads and uterus. GnRH agonists, such as deslorelin and leuprolide acetate (Lupron Depot; TAP Pharmaceuticals), act by overriding normal pulsatile GnRH activity that is necessary for ovarian cyclicity and testicular function, thereby suppressing the hypothalamic-pituitary-gonadal axis. Overall, GnRH agonists have not been associated with any significant side effects, although there have been concerns that ovarian activity will not necessarily resume after drug withdrawal. For example, the GnRH analogue deslorelin (Suprelorin; Peptech Animal Health), administered as an implant to domestic cats, effectively suppressed ovarian follicular activity for extended periods without untoward health effects except weight gain (Munson et al. 2002a). In cats, however, ovarian suppression persisted well beyond the estimated efficacy period of the drug, and 8 of 10 cats did not fully return to normal estrous cycling within the study period. Because reliable reversibility has not yet been proven, more extended studies to test reversibility are indicated. In males, GnRH agonists suppress testosterone production and can cause loss or changes in secondary sex characteristics, such as deformed antler growth in deer.

ANDROGENS

Mibolerone (Cheque Drops: Pharmacia and Upjohn) is an androgenic, nonprogestational steroid that blocks LH release through negative feedback, thereby suppressing gonad function (Plumb 2002). In a pilot study in gray wolves, increased intraspecific aggression occurred, resulting in termination of the study (Gardner et al. 1995). These behavioral effects will limit usefulness of this contraceptive in captive animals. Furthermore, the need for daily administration makes use in free-ranging animals impractical. Mane growth occurred in a female lion (Gardner et al. 1995), and masculinization of external genitalia occurred in juvenile and female fetuses of treated dogs (Plumb 2002). In cats, mibolerone is toxic to the liver at doses needed for contraception, so its use is contraindicated in felids.

IMMUNOCONTRACEPTIVES

Immunocontraceptive vaccines, such as PZP (native porcine zona pellucida), ZP3 (synthetic zona pellucida), SpayVac (TerraMar Environmental Research), GnRH,

LH, and sperm proteins, induce an immune response against endogenous antigens that regulate endocrine cyclicity or fertilization. Because these antigens are viewed by the immune system as "self," they have low antigenicity and need to be made more immunogenic (that is, viewed as foreign by the immune system) to be effective. Immunogenicity is increased by tagging the antigen to another protein, delivering them with a potent adjuvant that stimulates a nonspecific inflammatory response (Ndolo et al. 1996; Skinner et al. 1996), or delivering them with an infectious disease agent (Holland and Jackson 1994). Clinical trials with immunocontraceptives have used different adjuvants, different antigens, and different vaccination protocols, and this variation in methods hampers amalgamation of knowledge from these trials. Furthermore, the outcomes of the trials have been measured differently, and safety evaluation has not been a primary objective. Therefore, information on the safety of immunocontraceptives in captive and free-ranging wild animals is lacking.

The immunocontraceptive most commonly used is a native PZP vaccine. The antibodies are directed against the zona pellucida, the coating on developing and mature ova where the sperm receptors (ZP3) are located. Because the zona pellucida is present in the ovary, there have been concerns that immune-mediated damage to the ovary may result from inciting an immune response against these proteins. However, no damage to ovarian follicles has been noted in any PZP-vaccinated species submitted to the Center (including felids, cervids, and canids). Also, no ovarian damage was seen in mice vaccinated with ZP (Tung et al. 1996). However, other studies have reported oophoritis (inflammation of the ovary) in response to ZP vaccination in macaques (Bagavant et al. 2002) and in ZP3-immunized mice (Skinner et al. 1984). Reduced follicular development has been noted in dogs, rabbits, mice, and some primates immunized with ZP (Govind et al. 2002; Mahi-Brown et al. 1988, 1992), but in one study, it was not clear if these effects were caused by ZP antibody-mediated cytotoxicity or Freund's complete adjuvant (Upadhyay et al. 1989). Additionally, long-term immunization of baboons with ZP antigens resulted in increased follicular atresia in some animals (Dunbar et al. 1989). Further histopathological evaluations of ZP-contracepted animals are indicated to assess possible permanent ovarian damage. If irreversible damage to the ovary occurs, then immunocontraceptives are not appropriate for use in rare and endangered species or wildlife populations that need reversible reproductive control. However, for situations in which sterilization is acceptable or desirable, immunocontraceptives that damage the ovary may be an attractive alternative and more easily administered than other forms of permanent sterilization.

Selection of the adjuvant for use for PZP immunocontraception may also have unintended consequences. Strong antibody responses to immunocontraceptives occur when Freund's complete adjuvant is used to enhance the immune response. However, Freund's complete adjuvant incites a marked granulomatous reaction in some species (Broderson 1989; Yamanaka et al. 1992). When used in zoo felids, the inflammation associated with Freund-adjuvanted PZP vaccines resulted in extensive tissue damage, leading to fistulas and abscesses at the injection site in some animals (L. Munson, unpublished data). Some domestic cats vaccinated with PZP in Freund's adjuvant developed prolonged hypercalcemia (L. Munson, unpublished data). An additional concern is that Freund's complete adjuvant contains heat-killed *Mycobacterium tuberculosis* that will cause a positive reaction in immunocontracepted animals when performing intradermal tests for tuberculosis. This reaction can confound interpretation of tests conducted to detect active *Mycobacterium* infections in primates or ungulates (Kirkpatrick et al. 1995). These adverse effects have not been noted in ungulates vaccinated with PZP in incomplete Freund's adjuvant. Therefore, selection of an adjuvant should consider potential side effects.

Use of PZP immunocontraceptives in seasonal breeders may result in delayed breeding and birth of offspring during seasons that are less than optimal for neonatal survival (McShea et al. 1997). An additional concern with PZP immunocontraceptives is that, because they do not suppress ovarian activity, immunocontracepted animals can have recurrent or persistent estrus. In ungulates, persistent estrus may lead to weight loss (Fraker et al. 2002), whereas carnivores treated with immunocontraceptives would continue to be exposed to endogenous steroids with the same pathological effects as exogenous progestins. Another concern is that immunocontraceptives, which rely on a healthy immune system to achieve contraception, may not be effective in animals with compromised immune function. In theory, contraception would fail in immunocompromised animals, which would have a negative effect on natural selection for population fitness.

Other immunocontraceptives being developed include anti-GnRH and anti-LH vaccines. The immune response in GnRH-vaccinated cervids caused delayed antler growth and retention of velvet (Miller et al. 2000), similar to the effects seen in progestin- and GnRH analogue-treated animals, because antler growth depends on testosterone production, which in turn is dependent on pulsatile GnRH. Active immunization against GnRH in pigs also caused damage to cells in the hypothalamus other than those producing GnRH (an action called a bystander effect) (Molenaar et al. 1993), but similar safety studies have not been conducted in wild animals. Similar concerns surround the use of immunotoxins that target cells regulating reproductive function. As these contraceptives become available, safety trials should be performed to assure that nonreproductive functions of the hypothalamus and pituitary gland are retained in treated animals.

VAS PLUGS

Occlusive, preformed Silastic vas deferens plugs are reversible contraceptives designed for men (Chen et al. 1996). However, preformed plugs do not conform to the varying size of the vas deferens in different species (L. Zaneveld and C. Asa, personal communication). When soft Silastic was injected into vas deferens to form plugs that mold to each vas deferens size, local inflammatory reactions occurred at sites of injection as a result of exuded sperm, rendering the animals sterile. Otherwise, the Silastic implants were well tolerated with only minimal compressive changes to the epithelium of the vas. Because the local inflammatory reactions resulted in regional occlusion, this method has no advantage over vasectomy.

MALE CHEMICAL CONTRACEPTIVES

Compounds such as bisdiamine and gossypol that arrest spermatogenesis have promise as male contraceptives, but the potential for serious side effects has limited their use. Orally administered bisdiamines have proven safe in males, but these compounds cause serious teratogenic effects if ingested by pregnant females (Taleporos et al. 1978). Also, testosterone decreased in male cats receiving oral bisdiamine contraceptives (L. Munson, unpublished data), which would be of concern in species with testosterone-dependent secondary sex characteristics. Gossypol can cause hepatic and cardiac necrosis, hypokalemia, or thyroid damage (Jones et al. 1997; Kumar et al. 1997; Rikihisa and Lin 1989).

OTHER PROBLEMS ASSOCIATED WITH CONTRACEPTIVES

In addition to adverse hormonal effects of MGA implants, abscesses and seromas have developed at the implant site. The occurrence of abscesses can be minimized by using proper ethylene oxide sterilization and poststerilization degasing of implants, as well as by practicing sterile surgical techniques during placement. MGA implants also have migrated from the original site, impeding timely removal. If implant removal cannot occur or absorption is augmented by fragmentation, the animal may be exposed to higher doses that could exacerbate side effects (Black et al. 1979).

For long-acting contraceptives administered by depot injection, injection site reactions rarely occur. For example, a small number of California sea lions (*Zalophus californianus*) and California sea otters (*Enhydra lutris*) developed injection site discomfort, lameness, anorexia, and lethargy following depot leuprolide administration (Calle et al. 1997).

Although the focus of this chapter is potential adverse medical consequences of contraceptive application, there also may be negative social or behavioral consequences. Repetitive estrous cycles caused by induced infertility may result in an increased level of aggression between males in a group in which females are continually in estrus, or may cause increased chasing or following behavior that may also have deleterious health consequences. Lack of offspring in a group may also limit the behavioral experience of group members, especially the exposure of younger females to appropriate mothering behavior. Additionally, as mentioned previously, offspring may be born in less advantageous times of the year if contraceptive efficacy wanes during the last months of a breeding season.

Potential adverse effects of the contraceptive agents themselves are one concern in the selection of an appropriate contraceptive. There are also various clinical considerations related to administration method and other components of contraceptive use that need to be taken into account to select the optimal agent for a particular individual, species, institution, or management scheme. Anesthesia and surgery are required to accomplish some contraceptive and sterilization techniques (such as castration, vasectomy, tubal ligation, ovariohysterectomy, vas plugs, and MGA implants). For most species, anesthesia and surgery present minimal risks to accomplish these procedures. However, for some species (such as giraffes, elephants, hippopotamus, pinnipeds, and cetaceans), anesthesia presents greater challenges with potential for adverse consequences, so contraceptives that require these techniques are less desirable. Further, the anatomy and natural biology of some species make surgery more difficult and carry a higher risk of complications, for example, the dense blubber layer and totally aquatic life history of cetaceans. Therefore, selection of a contraceptive needs to consider the method of application in addition to other safety issues. These topics are covered in more depth in Chapter 6, Choosing the Most Appropriate Contraceptive.

SUMMARY

As with the administration of any drug or vaccine, adverse reactions will likely occur in a few individual animals given contraceptives, but overall the benefit of contraception should outweigh the risk of harm when selecting a contraceptive for a species. Most ungulates have no adverse reactions to either PZP vaccines or progestins. Also, safe contraceptives are currently available for most primate species, although MGA should be used with caution in Lemuridae, Goeldi's monkeys, and squirrel monkeys. In contrast, contraceptive safety in carnivores, particularly felids and canids, continues to be a problem. Because of efficacy, availability, and ease of administration, most zoos still rely on progestin contraceptives for carnivores despite the serious deleterious effects. GnRH agonists hold promise for carnivore contraception, but most forms are still expensive or difficult to procure.

Surveillance for immediate as well as long-term adverse effects of contraception will be continued by the Center. The Center requests reproductive tracts from all captive and free-ranging mammals so that the risks of contraception can be assessed and factored into ethical management of captive animals. Additional clinical trials assessing contraceptive safety should be conducted, particularly as bioengineered contraceptives become available. Domestic cats and dogs could serve as valuable models for wild carnivores in these trials.

CONTRIBUTING INFORMATION OR TISSUES TO THE CENTER

Animal managers in wildlife reserves or zoos should contact the Center if adverse reactions are suspected. Submission of reproductive tracts from all female captive and free-ranging wildlife mammals is encouraged, regardless of whether they were treated with contraceptives. Reproductive tracts from males treated with contraceptives should also be submitted. Protocols and forms for submission can be found at http://www.aazv.org/secure/cagpathologysurvey.htm or www.stlzoo.org/contraception.

REFERENCES

Bagavant, H., C. Sharp, B. Kurth, and T. S. K. Tung. 2002. Induction and immunohistology of autoimmune ovarian disease in cynomolgus macaques (*Macaca fascicularis*). *Am J Pathol* 160:141–149.

Black, D., U. S. Seal, E. D. Plotka, and H. Kitchen. 1979. Uterine biopsy of a lioness and tigress after melengestrol implant. *J Zoo Anim Med* 10:53–56.

Broderson, J. R. 1989. A retrospective review of lesions associated with the use of Freund's adjuvant. *Lab Anim Sci* 39:400–405.

Calle, P. P., M. D. Stetter, B. L. Raphael, R. A. Cook, J. Basinger, H. Walters, and K. Walsh. 1997. Use of depo-leuprolide acetate to control undesirable male associated behaviors in the California sea lion (*Zalophus californianus*) and California sea otter (*Enhydra lutris*). In *Proceedings, International Association of Aquatic Animal Medicine*, Vol. 28, 6–7, May 3–7, Harderwijk, The Netherlands.

Chassy, L. M., I. A. Gardner, E. D. Plotka, and L. Munson. 2002. Genital tract smooth muscle tumors are common in zoo felids but are not associated with melengestrol acetate contraceptive treatment. *Vet Pathol* 39:379–385.

Chen, Z., Y. Gu, X. Liang, L. Shen, and W. Zou. 1996. Morphological observations of vas deferens occlusion by the percutaneous injection of medical polyurethane. *Contraception* 53:275–279.

Clark, J. H., and S. K. Mani. 1994. Actions of ovarian steroid hormones. In *The Physiology of Reproduction*, ed. E. Knobil and J. D. Neill, 1011–1059. New York: Raven Press.

Dunbar, B. S., C. Lo, J. Powell, and V. C. Stevens. 1989. Use of a synthetic peptide adjuvant for immunization of baboons with denatured and deglysolated pig zona pellucida glycoproteins. *Fertil Steril* 52:311–318.

Food and Drug Administration. 1997. *Guidance for Industry: M3 Nonclinical Safety Studies for the Conduct of Human Clinical Trials for Pharmaceuticals*. Rockville, MD: US Dept. of Health and Human Services, Food and Drug Administration, Drug Information Branch, Center for Drug Evaluation Research, Center for Biologics Evaluation and Research. (http://www.fda.gov/cder/guidance/1855fnl.pdf)

Fraker, M. A., R. G. Brown, G. E. Gaunt, J. A. Kerr, and B. Pohajdak. 2002. Long-lasting, single-dose immunocontraception of feral fallow deer in British Columbia. *J Wildl Manag* 66:1141–1147.

Gardner, H. M., W. D. Hueston, and E. F. Donovan. 1995. Use of mibolerone in wolves and in three *Panthera* species. *J Am Vet Med Assoc* 187:1193–1194.

Govind, C. K., N. Srivastava, and S. K. Gupta. 2002. Evaluation of the immunocontraceptive potential of *Escherichia coli* expressed recombinant non-human primate zona pellucida glycoproteins in homologous animal model. *Vaccine* 21:78–88.

Hafez, E. S. E., M. R. Jainudeen, and Y. Rosnina. 2000. Hormones, growth factors, and reproduction. In *Reproduction in Farm Animals*, ed. E. S. E. Hafez and B. Hafez, 33–54. Philadelphia: Lippincott Williams & Wilkins.

Hansen, P. J. 2000. Immunology of reproduction. In *Reproduction in Farm Animals*, ed. E. S. E. Hafez and B. Hafez, 341–353. Philadelphia: Lippincott Williams & Wilkins.

Harrenstien, L. A., L. Munson, and U. S. Seal. The American Zoo and Aquarium Association Mammary Cancer Study Group. 1996. Mammary cancer in captive wild felids and risk factors for its development: a retrospective study of the clinical behavior of 31 cases. *J Zoo Wildl Med* 27:468–176.

Holland, M. K., and R. J. Jackson. 1994. Virus-vectored immunocontraception for control of wild rabbits: identification of target antigens and construction of recombinant viruses. *Reprod Fertil Dev* 6:631–642.

Jones, T. C., R. D. Hunt, and N. W. King. 1997. *Veterinary Pathology*. Baltimore: Williams & Wilkins.

Kazensky, C. A., L. Munson, and U. S. Seal. 1998. The effects of melengestrol acetate on the ovaries of captive wild felids. *J Zoo Wildl Med* 29:1–5.

Kirkpatrick, J. F., W. Zimmermann, L. Kolter, and J. W. Turner. 1995. Immunocontraception of captive exotic species. I. Przewalski's horses (*Equus przewalskii*) and Banteng (*Bos javanicus*). *Zoo Biol* 14:403–416.

Kollias, G. V., M. B. Calderwood-Mays, and B. G. Short. 1984. Diabetes mellitus and abdominal adenocarcinoma in a jaguar receiving megestrol acetate. *J Am Vet Med Assoc* 185:1383–1386.

Kumar, M., S. Sharma, and N. K. Lohiya. 1997. Gossypol-induced hypokalemia and the role of exogenous potassium salt supplementation when used as an antispermatogenic agent in male langur monkey. *Contraception* 56:251–256.

Lauderdale, J. W. 1983. Use of MGA (melengestrol acetate) in animal production. In *Anabolics in Animal Production*, ed. E. Meissonnier, 193–212. Office International des Epizooties, Paris.

Linnehan, R. M., and J. L. Edwards. 1991. Endometrial adenocarcinoma in a bengal tiger (*Panthera tigris bengalensis*) implanted with melengestrol acetate. *J Zoo Wildl Med* 22:130–134.

Mahi-Brown, C. A., R. Yanagimachi, M. L. Nelson, H. Yanagimachi, and N. Palumbo. 1988. Ovarian histopathology of bitches immunized with porcine zonae pellucidae. *Am J Reprod Immunol Microbiol* 18:94–103.

Mahi-Brown, C. A., R. P. McGuinness, and F. Moran. 1992. The cellular immune response to immunization with zona pellucida antigens. *J Reprod Immunol* 21: 29–46.

McShea, W. J., S. L. Monfort, S. Hakim, J. F. Kirkpatrick, I. Liu, L. M. Chassy, and L. Munson. 1997. The effect of immunocontraception on the behavior and reproduction of white-tailed deer. *J Wildl Manag* 61:560–569.

Miller, L. A., B. E. Johns, and G. J. Killian. 2000. Immunocontraception of white-tailed deer with GnRH vaccine. *Am J Reprod Immunol* 44:266–274.

Mol, J. A., E. van Garderen, G. R. Rutteman, and A. Rijnberk. 1996. New insights in the molecular mechanism of progestin-induced proliferation of mammary epithelium: induction of the local biosynthesis of growth hormone (GH) in the mammary gland of dogs, cats and humans. *J Steroid Biochem Mol Biol* 57:37–71.

Molenaar, G. J., C. Lugard-Kok, R. H. Meloen, R. B. Oonk, J. de Koning, and C. J. G. Wensing. 1993. Lesions in the hypothalamus after active immunization against GnRH in the pig. *J Neuroimmunol* 48:1–12.

Mossman, H. W. 1987. Comparative anatomy of the oviduct, uterus, and vagina. In *Vertebrate Fetal Membranes*, ed. H. W. Mossman, 63–73. New Brunswick: Rutgers University.

Munson, L., J. E. Bauman, C. Asa, M. Jöchle, and T. E. Trigg. 2002a. Efficacy of the GnRH analogue deslorelin for suppression of oestrus cycle in cats. *J Reprod Fertil Suppl* 57:269–273.

Munson, L., I. A. Gardner, R. J. Mason, L. M. Chassy, and U. S. Seal. 2002b. Endometrial hyperplasia and mineralization in zoo felids treated with melengestrol acetate contraceptives. *Vet Pathol* 39:419–427.

Murnane, R. D., J. M. Zdziarski, T. F. Walsh, M. J. Kinsel, T. P. Meehan, P. Kovarik, M. Briggs, S. A. Raverty, and L. G. Phillips. 1996. Melengestrol acetate-induced exuberant endometrial decidualization in Goeldi's marmoset (*Callimico goeldii*) and squirrel monkey (*Saimiri sciureus*). *J Zoo Wildl Med* 27:315–324.

Ndolo, T. M., M. Oguna, C. S. Bambra, B. S. Dunbar, and E. D. Schwoebel. 1996. Immunogenicity of zona pellucida vaccines. *J Reprod Fertil Suppl* 50:151–158.

Nettles, V. F. 1997. Potential consequences and problems with wildlife contraceptives. *Reprod Fertil Dev* 9:137–143.

O'Brien, T. D., P. C. Butler, P. Westermark, and K. H. Johnson. 1993. Islet amyloid polypeptide: a review of its biology and potential roles in the pathogenesis of diabetes mellitus. *Vet Pathol* 30:317–332.

O'Brien, T. D., J. D. Wagner, K. N. Litwak, C. S. Carlson, W. T. Cefalu, K. Jordan, K. H. Johnson, and P. C. Butler. 1996. Islet amyloid and islet amyloid polypeptide in cynomolgus macaques (*Macaca fascicularis*): an animal model of human non-insulin-dependent diabetes mellitus. *Vet Pathol* 33:479–485.

Orford, H. J. L., and R. Perrin. 1988. Contraception, reproduction and demography of free-ranging Etosha lions (*Panthera leo*). *J Zool* 216:717–733.

Plumb, D. C. 2002. *Veterinary Drug Handbook*. Ames: Iowa State Press.

Porton, I. 1995. Results for primates from the AZA contraception database: species, methods efficacy and reversals. In *Proceedings, Joint Conference of the American Association of Zoo Veterinarians, Wildlife Disease Association, and American Association of Wildlife Veterinarians*, 335–349, August 12–17, 1995, East Lansing, MI.

Portugal, M. M., and C. S. Asa. 1995. Effects of chronic melengestrol acetate contraceptive treatment on perineal tumescence, body weight, and sociosexual behavior of hamadryas baboons (*Papio hamadryas*). *Zoo Biol* 14:251–259.

Raphael, B. L., P. Kalk, P. Thomas, P. P. Calle, K. Dohi, and R. A. Cook. 2003. The use of melengestrol acetate in feed for contraception in herds of captive ungulates. *Zoo Biol* 22:455–463.

Rikihisa, Y., and Y. C. Lin. 1989. Effect of gossypol on the thyroid in young rats. *J Comp Pathol* 100:411–417.

Seal, U. S., R. Barton, L. Mather, K. Olberding, E. D. Plotka, and C. W. Gray. 1976. Hormonal contraception in captive female lions (*Panthera leo*). *J Zoo Anim Med* 7:12–20.

Skinner, S. M., T. Mills, H. J. Kirchick, and B. S. Dunbar. 1984. Immunization with zona pellucida proteins results in abnormal ovarian follicular differentiation and inhibition of gonadotropin-induced steroid secretion. *Endocrinology* 115:2418–2432.

Skinner, S. M., S. V. Prasad, T. M. Ndolo, and B. S. Dunbar. 1996. Zona pellucida antigens: targets for contraceptive vaccines. *Am J Reprod Immunol* 35:163–174.

Stock, M. K., and J. Metcalfe. 1994. Maternal physiology during gestation. In *The Physiology of Reproduction*, ed. E. Knobil and J. D. Neill, 947–983. New York: Raven Press.

Straub, S. G., G. W. Sharp, M. D. Meglasson, and C. J. De Souza. 2001. Progesterone inhibits insulin secretion by a membrane delimited, non-genomic action. *Biosci Rep* 21:653–666.

Taleporos, P., M. P. Salgo, and G. Oster. 1978. Teratogenic action of a bis(dichloroacetyl)diamine on rats: patterns of malformations produced in high incidence at time-limited periods of development. *Teratology* 18:5–16.

Tung, K. S. K., J. Ang, and Y. Lou. 1996. ZP3 peptide vaccine that induces antibody and reversible infertility without autoimmune oophritis. *Am J Reprod Immunol* 35: 181–183.

Turner, J. W., I. K. M. Liu, and J. F. Kirkpatrick. 1992. Remotely delivered immunocontraception in captive white-tailed deer. *J Wildl Manag* 56:154–157.

Upadhyay, S. N., P. Thillai-Koothan, A. Bamezai, S. Jayaraman, and G. P. Talwar. 1989. Role of adjuvants in inhibitory influence of immunization with porcine zona pellucida antigen (ZP-3) on ovarian folliculogenesis in bonnet monkeys: a morphological study. *Biol Reprod* 41:665–673.

White, L. M., R. J. Warren, and R. A. Fayrer-Hosken. 1994. Levonorgestrel implants as a contraceptive in captive white-tailed deer. *J Wildl Dis* 30:241–246.

Yamanaka, M., K. Hiramatsu, T. Hirahara, T. Okabe, M. Nakai, N. Sasaki, and M. Goto 1992. Pathological studies on local tissue reactions in guinea pigs and rats caused by different adjuvants. *J Vet Med Sci* 54:685–692.

CHERYL S. ASA, INGRID J. PORTON, AND PAUL P. CALLE

6
CHOOSING THE MOST APPROPRIATE CONTRACEPTIVE

The reproductive management plan for burgeoning populations of free-ranging wildlife typically calls for a reduction in the overall population growth rate. Less concern is directed at micromanaging the reproductive contribution of individual animals. In contrast, captive populations of mammals housed in zoos are small and therefore subject to a rapid reduction in the population's genetic diversity. Modern captive breeding programs that aim to reduce the loss of genetic diversity must, by necessity, manage the reproductive contribution of each individual. Ideally, captive management plans identify the number of offspring required from each individual, which in turn drives the selection of the most appropriate contraceptive method for each stage of the animal's life.

Such plans must also include the realization that captive populations are subject to chance events, however. Offspring or relatives may die before they have a chance to reproduce. Permanent sterilization of middle-aged adults is, therefore, typically undesirable. Factors that must be considered when selecting a contraceptive method include not only the products and delivery methods but also the species, age, condition, and circumstances of the animals to be treated and the consequences of unwanted pregnancies or sterility. Methods must be cost-effective to be practically utilized. Variables such as the costs of medication, administration method, need for anesthesia or surgery, and required investment of personnel and time all factor into the analysis. The method that is most appropriate for a given species or social group will vary between institutions for site-specific reasons.

NEED FOR CERTAINTY OF EFFICACY

For most captive wildlife, 100 percent efficacy is desired, limiting the choices to those methods that have already been proven effective for the species in question. To date, the most effective and most widely used contraceptive for captive wildlife is the melengestrol acetate (MGA) implant (available from E. D. Plotka). It has also been used for the longest time, more than 25 years in large felids and 15 to 20 years in primates, but for considerably less time in most other taxa (Contraception Database). Other methods have been tested much less thoroughly, and their efficacy varies rather dramatically. Porcine zona pellucida (PZP) is not recommended for carnivores and is somewhat questionable in primates, but it seems to work well in most ungulates (see Chapter 9, Contraception in Ungulates; Chapter 13, Contraception in Free-Ranging Wildlife). Gonadotropin-releasing hormone (GnRH) agonists seem to work in most carnivores (inadequate dosages may have explained the few failures), but these are ineffective in male ungulates (see Chapter 9, Contraception in Ungulates).

In contrast to captive animals, contraception in free-ranging populations may be considered successful even when levels of efficacy are not much above 50 or 60 percent. For this application, controlling reproduction in individual animals is less important than reducing the reproductive rate for the population in general. Thus, the level of efficacy will be determined by the models for population regulation.

ISSUES RELATING TO DELIVERY

A variable related to certainty of efficacy is certainty of delivery. When prevention of reproduction is critically important, delivery systems that cannot guarantee administration of the complete contraceptive dose would be contraindicated. Required effective drug dose may be a limiting factor for injectable medications if the volume exceeds the amount that can reasonably be delivered by dart or pole syringe. Treatment of large numbers of individual animals maintained in herds or groups can be challenging. Incomplete administration resulting from dart malfunction also can be problematic. Preparations such as PZP, depending upon the adjuvant used, may become quite viscous in cold weather, so that effective delivery may require a large-bore needle or a delay until weather conditions are more moderate. Because some species have thick skin (for example, tapir species, hippopotamus, and giraffe) or a dense layer of subcutaneous fat (for example, ursids, suids, hippopotamus, pinnipeds, and cetaceans), long needles are needed to effectively deliver the medication for intramuscular injection. Appropriate selection of dart, needle, and delivery system can overcome these obstacles.

Oral contraceptives may be difficult to administer effectively to either individuals or groups of animals. The altered appearance or taste of the medicated feed may result in less food consumption than expected and therefore poor compliance in consuming the intended effective dose. When medications are sprinkled on feed for the group, distribution may be variable throughout the ration and some individuals may not consistently consume the desired contraceptive dose. This disadvantage can be overcome when medication is incorporated into the feed to achieve an even distribution throughout the ration. However, in either case, because subordinate animals with a low social rank may not consume the desired quantity of medication each day, contraceptive failures may occur in these animals. Individual oral administration may also be challenging in species such as the great apes that may appear to consume an individually offered contraceptive; in reality, they may conceal the intended medication (such as a birth control pill) within the mouth and then reject it later.

Anesthesia and surgery are required for most permanent sterilization techniques (for example, castration, vasectomy, tubal ligation, and ovariohysterectomy) and to insert some implants. For most species, anesthesia and surgery present minimal risks, but for some (such as giraffes, hippopotamus, equids, pinnipeds, and cetaceans), anesthesia and surgery present greater challenges and potential for adverse consequences, making these techniques less desirable. The anatomy and biology of some species, such as the dense blubber layer and totally aquatic habits of cetaceans, also mean that contraceptive techniques that depend upon surgery are not optimal choices.

NEED FOR TIMED REVERSAL

In some cases, it may be determined that contraception is needed indefinitely, perhaps permanently, but sterilization is avoided because the circumstances may change such that reproduction is again appropriate. In these cases, it is only important that a method be reversible. However, in most instances, controlling the time of reversal is also important. The contraceptives with the most predictable reversal times are the oral preparations (for example, MGA in feed: Purina Mills, LLC; Regu-mate: Hoechst-Roussel) and removable implants (for example, MGA implants; Norplant: Wyeth-Ayerst), because, after cessation of feeding or implant removal, hormone levels fall below the level of efficacy within 1 to 2 days. Likewise, removal of intrauterine devices (IUDs) eliminates the block to initiation of pregnancy. However, depot injections (Lupron-Depot: TAP Pharmaceuticals; Depo-Provera; Pharmacia and Upjohn), deslorelin implants (Suprelorin: Peptech Animal Health), and PZP vaccines vary greatly in time to reversal (see Chapter 4, Assessing Efficacy and Reversibility). The variability in Depo-Provera reversibility is compounded by absorption into fat layers in obese individuals or in species with a dense layer of subcutaneous fat (such as ursids, suids, hippopotamus, pinnipeds, and cetaceans). Deslorelin implants (Suprelorin: Peptech Animal Health), in addition to having a variable time to reversal, are difficult to remove surgically because of their smaller size and biodegradation.

Some of the agents discussed as contraceptives can also be used to synchronize estrous cycles for natural breeding or as a part of artificially assisted reproductive strategies such as artificial insemination or embryo transfer. Timing of contraceptive application and withdrawal can also be effectively employed as a management strategy so that conceptions after reversal will result in offspring born during the most favorable times of the year (for example, planning births of outdoor ungulates to avoid winter months in temperate climates).

REPRODUCTIVE SEASONALITY

Choosing a contraceptive for seasonal breeders is in many ways easier. First, duration of efficacy can be less than 1 year, which may also attenuate any deleterious effects in species such as carnivores or lessen stress associated with repeated treatment. Second, preventing the resumption of reproductive function by commencing treatment during the nonbreeding season can be simpler than suppressing reproductive processes and may require lower doses. However, the length of the breeding season in contracepted animals may be longer than typical published values, because those values are usually for free-ranging populations of animals in which females have access to partners and become pregnant, ending ovulatory cycles. When pregnancy is prevented, females may continue to ovulate, as was observed in PZP-treated white-tailed deer (*Odocoileus virginianus*; McShea et al. 1997) and female black lemurs (*Eulemur macaco*) housed with vasectomized males (I. Porton, unpublished data).

AGE: PREPUBERTAL OR NEAR SENESCENCE

Because so few data are available concerning the use of contraception in prepubertal animals, additional care must be taken when choosing a method for young animals, especially those that are genetically valuable and expected to reproduce at a later date. Domestic cows whose mothers were fed MGA while pregnant and who themselves consumed MGA when young were reported to reproduce normally when they matured (Schul et al. 1970). Further evidence of the safety of MGA and a similar synthetic progestin, megestrol acetate, in prepubertal animals (Ovarid and Ovaban: Schering-Plough) comes from studies of male muntjac (Stover et al. 1987) and female dogs (Bigbee and Hennessy 1977). These progestins did not appear to affect future reproductive performance in either species. Similarly, prepubertal female ruffed lemurs (*Varecia variegata*) treated with progestins were fertile as adults (I. Porton, unpublished data), as were white-tailed deer treated with PZP vaccine (J. Kirkpatrick, personal communication). Although species differences could certainly occur, the results from these studies involving ungulates, a carnivore, and a primate suggest that progestins may not affect reproductive processes when administered before puberty. Fewer species treated with PZP have been evaluated, but preliminary observations also are encouraging.

In contrast, for older animals nearing the end of their reproductive life span, reversibility may not be important. Instead, health concerns such as the need to avoid the risk of anesthesia may dictate the choice of an oral or injectable preparation.

CONTRACEPTION DURING PREGNANCY

Although contraception is not needed during pregnancy, for logistical reasons it may be convenient or advantageous to initiate treatment before parturition. Especially in species that have a postpartum estrus soon after giving birth, it may be important to ensure that a subsequent pregnancy is prevented without undue disturbance to the mother and new baby. Unfortunately, many female-directed contraceptives can interfere either with the pregnancy itself (such as GnRH analogues) or with parturition (progestins with or without estrogens). Although progestins early in gestation may help maintain the pregnancy (Diskin and Niswender 1989), in some species they contribute to resorption, especially early in the gestation period (Shirley et al. 1995; Ballou 1996).

In general, progestins suppress contractility of uterine smooth muscle and so can inhibit parturition. However, some females contracepted with progestins have given birth without incident (Porton 1995; see Chapter 8, Contraception in Nonhuman Primates). Review of the literature and the Contraception Database suggests that several variables may affect the outcome of pregnancy for females receiving progestins: dosage (the dose required for contraceptive efficacy may be lower than that which inhibits parturition), the type of progestin (some may have more effect on the uterine myometrium or be more bioactive), and species differences (Zimbelman et al. 1970; Jarosz and Dukelow 1975; Plotka and Seal 1989; Shirley et al. 1995). However, data on PZP indicate that it may be the safest contraceptive method to use during pregnancy because it has no known effects on the uterus.

Estrogens may not interfere directly with pregnancy, but they are thought to affect fetal development. Early studies of the effects of contraceptives did not separate progestin-only from combination estrogen–progestin formulations. However, other and more recent assessments of currently available progestins have found no deleterious effects on embryo development: levonorgestrel (Norplant; Shirley and Bundren 1995), medroxyprogesterone acetate (Depo-Provera; Borgatta et al. 2002), and altrenogest (Regu-mate; Naden et al. 1990). Calves born to mothers fed MGA were smaller at birth, but no other deleterious effects were found, and growth to weaning was normal (Schul et al. 1970). Similarly, male and female domestic horses whose mothers were fed altrenogest during pregnancy showed normal growth and development as well as normal reproductive function as adults (Shoemaker et al. 1989; Squires et al. 1989).

CONTRACEPTION DURING LACTATION

For species that do not have lactational anestrus or amenorrhea, contraception may be needed during the nursing period. There are no known effects of PZP vaccine or GnRH agonists on milk production or on the growth and development of the infant, as these products are protein based and thus are broken down in the digestive process. However, progestins and estrogens, as steroid hormones, do appear in the milk of treated mothers and can be absorbed by a nursing infant. Because estrogens may interfere with milk production and affect the development of nursing young, they are not recommended for lactating females. In contrast, progestins do not interfere with, and in fact may slightly enhance, milk production and have not been found to have negative effects on growth or development of nursing infants (WHO 1994a, 1994b). Despite the lack of evidence for deleterious effects for progestins, to ensure safety, new forms of progestin-based contraceptives that are not orally active, and so cannot be absorbed by the infant, are being developed (for example, Nestorone; Massai et al. 2001).

EFFECTS ON SOCIAL GROUPS AND POPULATIONS

Contraception impacts not only the individual recipient of the contraceptive but also the social group in which the animal lives (for reviews, see Asa 1996a,

1996b). For example, depending on the method used, contraception of a female can cause the complete cessation of estrus or the occurrence of repetitive estrous cycles. Neither condition is "normal," because the typical condition would be regular pregnancies and periods of lactation, so the social consequences of altering the incidence of estrous cycles should be considered.

If contraception is needed for all the females in a social group of one male and several females, it would be simpler to treat only the male. In certain social groups, however, because continual estrous cycling can increase tension and aggression levels among the females, treating the females may be favored. In groups that contain multiple males, aggression that accompanies competition for estrous females might also argue for cycle suppression. If, on the other hand, interfemale aggression is not a problem and male contraception is preferred, it may be advantageous to select a contraceptive that suppresses testosterone as well as spermatogenesis.

Methods that do not suppress estrous behavior include tubal ligation and the PZP vaccine. Because PZP only interferes with fertilization, hormone production is not altered, except in cases where more general ovarian damage may occur (see Chapter 3, Types of Contraception). Another method that allows periodic estrous behavior is the combination birth control pill. Even though ovulation is suppressed, the pills are typically given for 3 weeks followed by 1 week of either placebo pills or nothing. The placebo week, which usually permits sufficient follicle growth and estrogen production to stimulate some signs of estrus, may also be accompanied by mating behavior. In some species, such as chimpanzees, learning plays a role in the acquisition of appropriate male copulatory behavior (King and Mellen 1994). In such cases, the complete suppression of mating behavior may not be advisable in social groups that contain maturing males.

In contrast, progestin-based contraceptives (such as MGA, Depo-Provera, Norplant) are administered continually with no withdrawal phase. Progestin treatment is most similar to pregnancy, and, as in pregnancy, estrous cycles are typically suppressed. However, because it is difficult to completely suppress follicle growth with progestins (see Chapter 3, Types of Contraception), estrous behavior may sometimes be expressed. Although progestins might be expected to affect social interactions, in at least the few studies that have been conducted no significant changes were observed either in captive or free-ranging social groups (hamadryas baboons, *Papio hamadryas*: Portugal and Asa 1995; Rodrigues fruit bats, *Pteropus rodricensis*: Hayes et al. 1996; lions, *Panthera leo*: Orford 1996; golden lion tamarins, *Leontopithecus rosalia*: Ballou 1996; and golden-headed lion tamarins, *Leontopithecus chrysomelas*: De Vleeschouwer et al. 2000).

Because GnRH agonists suppress the production of the sex steroids, their effect is most similar to ovariectomy in females and castration in males, with attendant elimination of estrous behavior and a reduction in male-typical behavior, respectively. However, as is also true of castration, male aggression and sexual interest may not be eliminated, especially in sexually experienced animals.

A possible behavioral result of contraception that has not been studied is the effect of the absence of young, both for individual females and for the social group, including peer socialization opportunities. In free-ranging animals, the potential for the breakdown of social bonds in contracepted pairs or groups as a result of the suppression of estrus or reproductive failure must be considered. The observation that pair-bonds appear to have endured in some wild canid species in which males were vasectomized indicates that, at least in the short term, production of young may not be necessary for pairs to remain together (gray wolves, *Canis lupus*: Mech et al. 1996; red foxes, *Vulpes vulpes*: Saunders et al. 2002). In addition to documenting maintenance of pair-bonds in a study of coyotes (*Canis latrans*), Bromley and Gese (2001) found that vasectomized males and tubally ligated females continued to maintain territories, a measure of social structure.

A further problem with implementation of fertility control, whether by reversible contraception or permanent sterilization, is achieving population regulation or reduction. It seems elementary to expect that preventing reproduction in even part of a population should reduce the number of individuals in the population. Unfortunately for wildlife managers, that is seldom the case. One obstacle is in applying the method to a sufficient number of individuals to achieve the management goal, because it is unlikely that 100 percent of the target animals (using either male- or female-directed methods) can be successfully treated. This limitation makes it unlikely that contraception can actually reduce population numbers for most species (Garrott 1991).

Most analyses have demonstrated that some proportion of animals must be killed to achieve reduction (Hone 1992; Bomford and O'Brien 1997). However, species with high age-specific mortality (population turnover rate, which is more common in short-lived, fast-reproducing species such as rodents) may be the exception. Because their natural mortality rates are high, limiting reproduction in such species may indeed decrease the population (Stenseth 1981). Although rabbits might be thought to conform to these specifications, Twigg et al. (2002) found that more than 90 percent of the study population had to be sterilized before even a marginal population decrease was detected. In contrast, contraception may be effective in ultimately reducing population numbers in longer-lived species, although over a considerably longer period of time, by using a method that persists for years and applying it to a large proportion of the population, especially if young animals are treated as they reach reproductive age (Stenseth 1981).

Clearly, knowledge of the species natural history and local ecology is critical to making such predictions. In fact, others have found that even when contraception or sterilization is effective in controlling reproduction in treated individuals, compensatory responses may negate the effect through population destabilization (Bomford and O'Brien 1997). For example, there may be increased survival of juveniles resulting from decreased competition (fewer other mouths to compete with), increased birth rates in untreated individuals from increased resource availability (for example, decreased competition for good territories), as well as increased immigration or decreased dispersal, both the result of more space being available because of the lower birth rate). Because juvenile mortality is high for many species, such compensatory mechanisms might be expected to be common. In fact, these responses are not restricted to predictions about contraceptive effects but have also been reported for recovery following culling (feral horses: Eberhardt et al. 1982; feral donkeys: Choquenot 1990).

Another aspect of a species social system that can confound efforts to control population numbers despite successful fertility control involves aspects of dominance and territoriality. If the method alters social behavior (e.g., by changing hormone levels), dominance may be lost, and otherwise nonreproductive, subordinate animals may begin to reproduce, resulting in no net change in population numbers. Likewise, if the drive to maintain a territory diminishes, the individual will be replaced by another reproductive individual (Porter et al. 1991). The implications for territorial males are obvious, but it is also necessary to ascertain whether hormonally normal females or only pregnant females maintain territories (Caughley et al. 1992; Davis and Pech 2002).

In contrast, strategies that target dominant males for sterilization, even if testosterone levels are not altered, risk causing another type of destabilization. Removing individuals that would otherwise be the primary breeders, or that would be most likely selected by females, has genetic consequences for the population, a situation referred to as the "fallacy of the superfluous male" (Willson 2002). The genetic consequences of artificial control of free-ranging populations have raised concerns, especially in regard to methods that cause permanent sterility. However, in most cases, target species are abundant and not in danger of extinction, so that even if local populations might crash as a result of disease or other unpredictable factor, the survival of the species would not be threatened. Thus, unless a population can be shown to be genetically unique, such concerns are probably unnecessary (Seal 1991).

Finally, an unexpected outcome of limiting lifetime reproductive output for females is the possibility for increased longevity, as has been found in feral horses treated with PZP vaccine (Turner and Kirkpatrick 2002). Mortality rates decreased and body condition scores increased in treated but not in control mares, and more treated mares reached age classes above 21 years. Comparable data are not available for other species except humans. A fascinating retrospective study of British aristocratic families, with records dating back to 740 A.D., revealed an inverse correlation between number of children and life span (Weestendorp and Kirkwood 1998). The effect was especially pronounced for women who reached menopause (that is, did not die young in childbirth or from infectious diseases that are independent of age). Because this cohort of women was unlikely to have experienced social or economic deprivation, the effects of reproduction on life span are more obvious. Thus, contraception may contribute to the overall health of females, which has implications for the age and class structures of populations as well as their absolute numbers.

However, the opposite effect is predicted for males in populations where females cannot conceive but still come into estrus and, because they do not become pregnant, will continue to cycle. Males would be expected to continue competing for and mating with these females, perhaps jeopardizing their own health as a consequence of the increased level of activity and possible fights, combined with less time available for resting and foraging or hunting, with the continued attempts to breed. Even without an extended breeding season, many male ungulates lose too much weight during the fall rut to survive winter (Lincoln et al. 1970). Thus, there may be differential effects on longevity by gender as well as by species.

PUTTING RISK IN PERSPECTIVE

All contraceptives available today carry some degree of risk, which must be weighed primarily against the risks associated with pregnancy and parturition (see Chapter 5, Adverse Effects of Contraceptives), the negative social consequences of separation or isolation, and the production of surplus animals (see Chapter 1, The Ethics of Wildlife Contraception). Very young or old females or those with compromised health are more likely to suffer complications from pregnancy and giving birth. In addition, the deleterious genetic effects associated with the production of young from mating with close relatives within a family group may well outweigh the risk of side effects from a contraceptive. The social and psychological benefits of maintaining individuals in family or other heterosexual groupings may be more important than avoiding potential health problems. Careful consideration of all available methods, especially in regard to their record for the species in question, can help minimize possible risks.

REFERENCES

Asa, C. S. 1996a. The effects of contraceptives on behavior. In *Contraception in Wildlife,* Book 1, ed. U. S. Seal, E. D. Plotka, and P. N. Cohn, 157–170. Lewiston, NY: Edwin Mellen.

Asa, C. S. 1996b. Physiological and social aspects of reproduction of the wolf and their implications for contraception. In *Ecology and Conservation of Wolves in a Changing World,* ed. L. Carbyn, S. H. Fritts, and D. R. Seip, 283–286. Edmonton: Canadian Circumpolar Institute.

Ballou, J. D. 1996. Small population management: contraception of golden lion tamarins. In *Contraception in Wildlife,* Book 1, ed. U. S. Seal, E. D. Plotka, and P. N. Cohn, 339–358. Lewiston, NY: Edwin Mellen.

Bigbee, H. G., and P. W. Hennessy. 1977. Megestrol acetate for postponing estrus in first heat bitches. *Vet Med Small Anim Clin* 72:1727–1730.

Bomford, M., and P. O'Brien. 1997. Potential use of contraception for managing wildlife pests in Australia. In *Contraception in Wildlife Management,* ed. T. J. Kreege, 205–215. USDA/APHIS Technical Bulletin No. 1853. Washington, DC: US Dept. of Agriculture.

Borgatta, L., A. Murthy, C. Chuang, L. Beardsley, and M. S. Burnhill. 2002. Pregnancies diagnosed during Depo-Provera use. *Contraception* 66:169–172.

Bromley, D., and E. M. Gese. 2001. Effects of sterilization on territory fidelity and maintenance, pair bonds, and survival rates of free-ranging coyotes. *Can J Zool* 79:386–392.

Caughley, G., R. Pech, and D. Grice. 1992. Effect of fertility control on a population's productivity. *Wildl Res* 19:623–627.

Choquenot, D. 1990. Rate of increase for populations of feral donkeys in northern Australia. *J Mammal* 71:151–155.

Davis, S. A., and R. P. Pech. 2002. Dependence of population response to fertility control on the survival of sterile animals and their role in regulation. *Reproduction Suppl* 60:89–103.

De Vleeschouwer, K., L. Van Elsacker, M. Heistermann, and K. Leus. 2000. An evaluation of the suitability of contraceptive methods in golden-headed lion tamarins (*Leontopithecus chrysomelas*), with emphasis on melengestrol acetate (MGA) implants. (II) Endocrinological and behavioural effects. *Anim Welf* 9:385–401.

Diskin, M. G., and G. D. Niswender. 1989. Effect of progesterone supplementation on pregnancy and embryo survival in ewes. *J Am Sci* 67:1559–1563.

Eberhardt, L. L., A. K. Majorowicz, and J. A. Wilcox. 1982. Apparent rates of increase for two feral horse herds. *J Wildl Manag* 46:367–374.

Garrott, R. A. 1991. Feral horse fertility control: potential and limitations. *Wildl Soc Bull* 19:52–58.

Hayes, K. T., A. T. C. Feistner, and E. C. Halliwell. 1996. The effect of contraceptive implants on the behavior of female Rodrigues fruit bats, *Pteropus rodricensis*. *Zoo Biol* 15:21–36.

Hone, J. 1992. Rate of increase and fertility control. *J Appl Ecol* 29:695–698.

Jarosz, S. J., and W. R. Dukelow. 1975. Effect of progesterone and medroxyprogesterone acetate on pregnancy length. *Lab Anim Sci* 35:156–158.

King, N. E., and J. D. Mellen. 1994. The effects of early experience on adult copulatory behavior in zoo-born chimpanzees (*Pan troglodytes*). *Zoo Biol* 13:51–59.

Lincoln, G. A., R. W. Youngson, and R. V. Short. 1970. The social and sexual behaviour of the red deer stag. *J Reprod Fertil Suppl* 11:71–103.

Massai, M. R., S. Diaz, E. Quinteros, M. V. Reyes, C. Herreros, A. Zepeda, H. B. Croxatto, and A. J. Moo-Young. 2001. Contraceptive efficacy and clinical performance of Nestorone implants in postpartum women. *Contraception* 64:369–376.

McShea, W. J., S. L. Monfort, S. Hakim, J. F. Kirkpatrick, I. K. M. Liu, J. W. Turner Jr., L. Chassy, and L. Munson. 1997. The effect of immunocontraception on the behavior and reproduction of white-tailed deer. *J Wildl Manag* 61:560–569.

Mech, L. D., S. H. Fritts, and M. E. Nelson. 1996. Wolf management in the 21st century: from public input to sterilization. *J Wildl Res* 1:195–198.

Naden, J., E. L. Squires, and T. M. Nett. 1990. Effect of maternal treatment with altrenogest on age at puberty, hormone concentrations, pituitary response to exogenous GnRH, oestrous cycle characteristics and fertility of fillies. *J Reprod Fertil* 88:185–195.

Orford, H. J. L. 1996. Hormonal contraception in free-ranging lions (*Panthera leo* L.) at the Etosha National Park. In *Contraception in Wildlife*, Book 1, ed. U. S. Seal, E. D. Plotka, and P. N. Cohn, 303–320. Lewiston, NY: Edwin Mellen.

Plotka, E. D., and U. S. Seal. 1989. Fertility control in deer. *J Wildl Dis* 25:643–646.

Porter, W. F., N. E. Matthews, H. B. Underwood, R. W. Sage, and D. F. Behrand. 1991. Social organization in deer: implications for localized management. *Environ Manag* 15:809–814.

Porton, I. 1995. Results for primates from the AZA contraception database: species, methods, efficacy and reversals. In *Proceedings, Joint Conference of the American Association of Zoo Veterinarians, Wildlife Disease Association, and American Association of Wildlife Veterinarians*, 381–394, August 12–17, 1995, East Lansing, MI.

Portugal, M. M., and C. S. Asa. 1995. Effects of chronic melengestrol acetate contraceptive treatment on perineal tumescence, body weight, and sociosexual behavior of hamadryas baboons (*Papio hamadryas*). *Zoo Biol* 14:251–259.

Saunders, G., J. McIlroy, M. Berghout, B. Kay, E. Gifford, R. Perry, and R. van de Ven. 2002. The effects of induced sterility on the territorial behaviour and survival of foxes. *J Appl Ecol* 39:56–66.

Schul, G. A., L. W. Smith, L. S. Goyings, and R. G. Zimbelman. 1970. Effects of oral melengestrol acetate (MGA®) on the pregnant heifer and on her resultant offspring. *J Anim Sci* 30:433–437.

Seal, U. S. 1991. Fertility control as a tool for regulating captive and free-ranging wildlife populations. *J Zoo Wildl Med* 22:1–5.

Shirley, B., and J. C. Bundren. 1995. Effects of levonorgestrel on capacity of mouse oocytes for fertilization and development. *Contraception* 51:209–214.

Shirley, B., J. C. Bundren, and S. McKinney. 1995. Levonorgestrel as a post-coital contraceptive. *Contraception* 52:277–281.

Shoemaker, C. F., E. L. Squires, and R. K. Shideler. 1989. Safety of altrenogest in pregnant mares and on health and development of offspring. *Equine Vet Sci* 9:69–72.

Squires, E. L., R. K. Shideler, and A. O. McKinnon. 1989. Reproductive performance of offspring from mares administered altrenogest during gestation. *Equine Vet Sci* 9: 73–76.

Stenseth, N. C. 1981. How to control pest species: application of models from the theory of biogeography in formulating pest control strategies. *J Appl Ecol* 18:773–794.

Stover, J., R. Warren, and P. Kalk. 1987. Effect of melengestrol acetate on male muntjac (*Muntiacus reevesi*). In *Proceedings, First International Conference of Zoological and Avian Medicine*, 387–388, September 6–11, Turtle Bay, HI.

Turner, A., and J. E. Kirkpatrick. 2002. Effects of immunocontraception on population, longevity and body condition in wild mares (*Equus caballus*). *Reproduction Suppl* 60:187–195.

Twigg, L. E., T. J. Lowe, G. R. Martin, A. G. Wheeler, G. S. Gray, S. L. Griffin, C. M. O'Reilly, D. J. Robinson, and P. H. Hubach. 2002. Effects of surgically imposed sterility on free-ranging rabbit populations. *J Appl Ecol* 37:16–39.

Weestendorp, R. G. J., and T. B. L. Kirkwood. 1998. Human longevity at the cost of reproductive success. *Nature* (Lond) 396:743–746.

Willson, M. F. 2002. The fallacy of the superfluous male. *Conserv Biol* 16:557–559.

World Health Organization (WHO) Task Force for Epidemiological Research on Reproductive Health. 1994a. Progestogen-only contraceptives during lactation. I. Infant growth. *Contraception* 50:35–54.

World Health Organization (WHO) Task Force for Epidemiological Research on Reproductive Health. 1994b. Progestogen-only contraceptives during lactation. II. Infant development. *Contraception* 50:55–68.

Zimbelman, R. G., J. W. Lauderdale, J. H. Sokoloski, and T. G. Schalk. 1970. Safety and pharmacologic evaluations of melengestrol acetate in cattle and other animals: a review. *J Am Vet Med Assoc* 157:1528–1536.

Part III

THE APPLICATION

Contraceptive methods that are used in captive or free-ranging wildlife are generally considered experimental, because, despite decades of use, most contraceptives have not been adequately monitored and evaluated for proper dosage, efficacy, safety, or reversal in the hundreds of mammalian species for which they are required. The American Zoo and Aquarium Association (AZA) Contraception Database was created so that information on the use of contraceptives in AZA institutions could be accumulated and evaluated to begin addressing that deficiency.

The chapters in Part III summarize the accumulated experience that has been gained using contraception to control reproduction in exotic mammals. The first five chapters (7 through 11) deal exclusively with captive mammals and provide a complete summary of the information collected through the AZA Contraception Database. The chapters also include a literature review of contraceptive methods that are commercially available or widely used in captive exotic mammals. Chapter 12 discusses the use of contraception in captive mammals, but in this case as a tool to address behavioral problems, specifically aggression. Chapter 13 addresses the use of contraception in free-ranging mammals and provides a framework for the issues and difficulties associated with wildlife fertility control. The authors point out that, despite substantial research directed at developing wildlife contraceptive methods, experience with applying these techniques in the field is still surprisingly limited.

Chapters 7 through 11 are organized along broad taxonomic divisions. The two mammal groups with the longest history of contraceptive use are carnivores and primates, which is reflected in the amount of the data presented in Chapters 7 (Contraception in Carnivores) and 8 (Contraception in Nonhuman Primates). That contraceptive use in exotic hoofed mammals (Chapter 9) and in cetaceans and pinnipeds (Chapter 10) is still relatively new is reflected in the smaller sample sizes reported in these chapters. Chapter 11 reviews what little information is available for the Hyracoidea, Scandentia, Rodentia, Edentata, Chiroptera, and Marsupialia, collectively referred to as "other mammals."

These taxon-based chapters provide a summary of what we know to date. They are not, however, intended to be complete how-to manuals: For this type of information, the reader is referred to Chapter 3, Types of Contraception, and to the AZA Contraception Advisory Group (CAG) Recommendations found at www.stlzoo.org/contraception. Furthermore, because the Recommendations are updated annually, based on an annual review of survey and research data, the reader can obtain the most current information from the Web site.

AZA CONTRACEPTION DATABASE

The Contraception Database encompasses the responses from the annual contraception surveys distributed by the AZA CAG. The surveys, sent to zoos and other facilities holding captive wildlife, request information on all the reversible methods of contraception that the facilities have used to treat mammals within their collection. The Contraception Database is not, however, exhaustive and comprehensive, because (1) not all surveyed institutions respond and (2) incomplete or missing records result in lost information. The latter is particularly true for historical records and for animals that have been removed from a zoo's collection and lost to follow-up. In addition, the data we have for international zoos are substantially less complete than those for AZA institutions because of the history of the database. The survey was initially distributed exclusively to AZA institutions, but later distribution was expanded to all national and international institutions that used melengestrol acetate (MGA) implants. Thus, in 1989 the first contraception survey was sent to 105 North American zoos whereas the 2003 mailing list included more than 500 facilities. Although the international institutions are only required to report on MGA implant use [receipt of MGA implants is dependent on responding to the survey, because the information is required in the United States for Food and Drug Administration (FDA) reports], they have increasingly provided data on other contraceptive methods. Nevertheless, there are hundreds of international zoos that are not surveyed.

DATA COLLECTION METHODS

The Contraception Database Survey requests information for all reversible contraceptive methods used in mammals. Data collected include species, sex, individual animal ID, previous births, contraceptive method, dose, date initiated, date of access to reproductive partner, date and reason for ending contraception, date of planned or unplanned parturition, and comments on behavioral or physical effects. All data are entered into a database program developed by the CAG. Annual surveys include a blank questionnaire for reporting new contraceptive bouts and a computer-generated survey for follow-up data on all ongoing contraceptive bouts. For example, if a female was treated with an MGA implant in 2001, the 2002 printout would include the data provided previously and would request information regarding whether the implant was still in place, had been removed, etc. If, on the other hand, an MGA implant has been removed to allow reproduction, the annual survey would ask whether the female had given birth. Data on permanent sterilization are not requested unless it is the method chosen to replace a reversible method of contraception.

TERMINOLOGY

A number of terms are used throughout the taxon chapters. Rather than repeatedly define the terms within each chapter, we offer the definitions here:

Contraceptive bout: The time period between initiation and termination of a method. For example, an MGA implant bout begins when an implant is inserted and ends when the implant is removed, whereas a contraceptive bout for an injectable contraceptive method such as Depo-Provera or porcine zona pellucida (PZP) is the time from injection through the period of estimated efficacy.

Failure: A failure can only occur when the contraceptive method is known to be in use and is within the period of estimated efficacy. For example, an MGA implant failure occurs when the implant is known to be in place for 2 years or less when conception occurs. If the implant is lost and the female conceives thereafter, the conception did not result from a dose or contraceptive drug failure but rather because the method was not in use. If the implant is in place for more than 2 years and the female conceives, we do not consider this a failure, but rather the information helps define the duration of efficacy.

Reversal: A reversal occurs when a female or male resumes fertility, which can be determined through a variety of methods (see Chapter 4, Assessing Efficacy and Reversibility). However, unless otherwise stated, in these chapters a successful reversal is defined as the first birth that occurs after termination of a contraceptive method. An abortion, stillbirth, and live birth are all considered reversals because they indicate that conception did occur. Reversal is defined in this manner because few zoos consistently monitor resumption of fertility through other techniques (e.g., urine or fecal steroids). On the other hand, zoos consistently record the occurrence of births or abortions, if detected.

Beyond the limited number of animal records per treatment, another factor that confounds interpretation of the Contraception Database is the variability in dosages. Not only are different contraceptives used but dosages are not calculated in the same way for each method. Even among the progestin-based contraceptives, different relative bioactivities of the various synthetic hormones preclude direct comparisons of dosage. For MGA implants, Dr. Plotka formulates each implant based on previously successful contraceptive dosages for the particular species by body weight, which is why species and body weight data must be submitted when placing an order. Each implant is tailored to the animal, and a continuous range of sizes can be produced. Each implant contains 20% MGA by weight, with 80% of the weight as Silastic. For new species, results from the most closely related species must be extrapolated. When sufficient data have been accumulated for calculation of minimum effective dosages, the results are provided to Dr. Plotka.

A somewhat similar situation exists for choosing the correct size of a deslorelin implant. Implants can be ordered through the AZA Wildlife Contraception Center, but only two sizes are currently being produced: 5 and 10 mg. At present, institutions requesting deslorelin for animals smaller than 30 kg are given 5-mg implants; the 10-mg size is provided to those treating larger animals.

Norplant implants (Wyeth-Ayerst) can be purchased commercially with a prescription, which gives the veterinarian some latitude in choosing a dose. The Norplant System available in the United States includes six silicone (Silastic) rods each containing 36 mg levonorgestrel, which totals 216 mg, the dosage prescribed for women. Norplant II (available in other countries but not in the United States) contains two 70-mg implants. The separate implants provide the veterinarian with the option of using between one and six rods, so that smaller species can be accommodated. However, the implants cannot be cut to produce smaller doses than these without compromising release rates.

Dosages for MGA in feed (ADF-16 Herbivore Diet: Purina Mills, LLC) are not calculated as milligrams of MGA per kilograms of body weight but as milligrams per animal per day, extrapolating from decades of data from domestic cattle. Only general categories of body size are considered, as for deslorelin implants. The AZA Wildlife Contraception Center nutrition consultant advises participating zoos on the ratios of medicated feed to use in the total diet.

The only dose–response study for Depo-Provera with an exotic species involved black lemurs (*Eulemur macaco*) (I. Porton and C. Asa, unpublished data). Because the results were similar to those for humans (efficacy at about 2.5 mg/kg body weight), a dose range of 2.5 to 5 mg/kg body weight has been recommended for mammals in general, with only a few exceptions. Although this dosage has been effective in most species, there are no data ascertaining whether this actually approximates the minimum effective dose for species other than the black lemur. An exception in efficacy of this dose range has been identified in New World monkeys, which have been shown to require considerably higher doses (at least 20 mg/kg body weight). In contrast, very large species such as the giraffe and hippopotamus have been successfully treated with substantially lower doses per kilogram of body weight. However, because exact body weights usually are not known for those species, the dosages are expressed simply as milligrams per injection estimated to be effective for a particular interval, based primarily on trial and error.

The other confounding variable in evaluating an effective dose for Depo-Provera is the interval between injections. For humans, injections are recommended every 3 months, but some animals have shown estrous behavior or even conceived within a shorter interval. Rather than increase the amount administered, some veterinarians or managers choose to administer the same amount but at more frequent intervals, complicating dosage comparisons. The same result could be achieved by increasing the amount given at the original intervals. What is required is keeping the circulating concentration of the synthetic hormone above the "threshold of efficacy," or the minimum effective concentration.

The time course of hormone release from the injection site into the circulation is bell shaped, starting at zero, increasing to a maximum, then tailing off. The maximum concentration reached is higher following injection of larger amounts. Within the first few days the blood concentration rises above the effective concentration, and sometime later it drops below this level. The time when it drops below the effective concentration depends on the maximum reached, or how much was injected. That is, it takes longer for higher doses to be metabolized and excreted, so higher doses remain effective for longer periods. However, there is also a range of safety; if too much is administered so that the maximum concentrations rise significantly above the threshold of efficacy, side effects could result. Thus, doses that achieve contraceptive efficacy for longer periods of time are increasingly more likely to be associated with deleterious effects.

For PZP, supplied by the Science and Conservation Biology Department at ZooMontana, the standard dose is 65 to 100 μg per animal, with the exception of elephants. Both 400 and 600 μg per animal have been successful in preventing pregnancies in elephants, and 200 μg is being tested to determine whether a lower dose would be adequate (J. Kirkpatrick, personal communication).

Birth control pills are used almost exclusively in great apes, the nonhuman primate taxon that is more similar in body weight to human females than are gibbons, monkeys, and prosimians. More than 50 combination birth control pill formulations have been developed, primarily to address the differences in unwanted side effects (e.g., hair growth, acne, weight gain, and depression) experienced by human females. Therefore, although no particular formulation of combination birth control pills has been shown to be ineffective in apes, the caretaking staff should be alert to individual differences in efficacy or side effects and, in consultation with a gynecologist, change the prescription accordingly.

Megestrol acetate pills (Ovaban and Ovarid: Schering-Plough) come in two sizes, 5 and 20 mg. The dosage for dogs in early proestrus is stated as 1 mg/lb body weight (2.2 mg/kg) in the United States and as 2 mg/kg body weight in the UK. Daily administration for 8 days is recommended to postpone estrus for 4 to 6 months. For dogs in anestrus, 32 days (US) or 40 days (UK) of treatment with 0.25 mg/lb (US) or 0.55 mg/kg (UK) of body weight is recommended. Administration of 0.2 mg/kg twice per week may be continued for 4 more months, but the manufacturer advises against longer continuous treatment or repeated treatment within the same year. When megestrol acetate tablets are used in captive animals, dosage is extrapolated from the dog data, and the drug is typically continued throughout the breeding season. Ovarid is approved for cats in the UK at 5 mg/cat per day for 3 days, then at 2.5 to 5 mg once weekly for 10 weeks to 18 months, but it is not approved for cats in the United States.

Another oral preparation, Regu-Mate (altrenogest in oil: Hoechst-Roussel), which is approved for horses, has been used in other equids and in some marine mammals. The solution at a dosage of 0.044 mg/kg body weight can be administered directly into the animal's mouth or placed on food.

Despite the more than 17,000 contraceptive records for 267 mammalian species, and despite more than 13 years of data collection through the Contraception Database and through Drs. Seal and Plotka for MGA dating back to the 1970s, the sample size per method for each species is astonishingly low and the number of reversal attempts per species is even lower. Yet the results concerning efficacy, at least for MGA, for the majority of species are encouraging, in many cases approaching 100 percent. However, in examining the Database it is clear that much more information is needed before we can be assured of the safety and reversibility as well.

KAREN E. DEMATTEO

7
CONTRACEPTION IN CARNIVORES

The problem resulting from large-bodied, litter-producing carnivores that breed well in captivity was recognized by Dr. Ulysses Seal in the early 1970s. Reproduction could be curtailed through sterilization, but such nonreversible options reduced the flexibility required to genetically and demographically manage breeding programs for endangered species. To address the need for reversible contraception, Seal et al. (1976) examined the use of synthetic progestins to achieve reversible contraception in exotic felids. In that study, medroxyprogesterone acetate (MPA) injections (Depo-Provera; Pharmacia and Upjohn) and implants were tested and compared to implants containing a different progestin, melengestrol acetate (MGA). Research trials with lions found that MPA and MGA were equally effective; however, MGA implants were preferred because of the ease of implant administration and removal, fewer effects on adrenocortical hormones, and minimal behavioral side effects. MGA implant trials were expanded to other felids including tigers (*Panthera tigris*), jaguars (*P. onca*), and leopards (*P. pardus*) and were also found to be successful (Seal et al. 1976). Since that time, hundreds of felids and many other mammals have been treated with MGA implants (now supplied by E. D. Plotka) and other contraceptive methods as a means of managing population growth. Here I review what has been learned on the subject of reversible contraception for carnivores.

This chapter summarizes the data that have been accumulated in the American Zoo and Aquarium Association (AZA) Contraception Advisory Group's (CAG) Contraception Database on the types of contraception that have been used in zoo-housed carnivores. Details concerning the Contraception Database and the data collection methods can be found in the introduction to Part III. In addition, the

Science and Conservation Biology (SCB) Department at ZooMontana formulates and distributes porcine zona pellucida vaccine (PZP) for use in captive and free-ranging wildlife. The data presented here on PZP use in captive carnivores are taken from their PZP Database.

The order Carnivora is composed of 7 families, and all are represented in the Contraception Database. However, most species and individuals are from the family Felidae. The number of taxa included are Felidae (13 genera/22 species), Canidae (9 genera/15 species), Ursidae (4 genera/7 species), Procyonidae (3 genera/4 species), Mustelidae (9 genera/9 species), Viverridae (6 genera/6 species), and Hyaenidae (3 genera/3 species) (see Appendix). The 15 methods of reversible contraception that have been used in carnivores include 2 progestin implants, 2 forms of injectable progestins, 6 orally delivered progestins, 3 gonadotropin-releasing hormone (GnRH) agonists, 1 immunocontraceptive, and 1 mechanical method (Table 7.1).

METHODS FOR FEMALES

Steroid Hormones

PROGESTINS: MELENGESTROL ACETATE IMPLANTS By far, the most frequently used method of contraception in zoo carnivores has been the MGA implant (see Table 7.1). Of the 2,590 MGA implants reported used in carnivores, the majority, 80 percent, were used in Felidae, another 8 percent in Canidae, 9 percent in Ursidae, 2 percent in Mustelidae, and less than 1 percent each in Procyonidae, Viverridae, and Hyaenidae. The estimated duration of efficacy for an MGA implant is 2 years (AZA CAG 2004), but the duration of completed MGA implant treatments with available end dates (that is, not lost to follow-up) varied, with 23 percent in use for less than 1 year, 28 percent for 1 to 2 years, 28 percent for 2 to 3 years, and 21 percent for more than 3 years.

MGA implants are highly effective in carnivores. The calculated efficacy for 1,430 MGA implants in use for 24 months was 99 percent. There were only 17 MGA implant failures, 16 in Felidae (less than 1 percent of total treatments) and 1 in Ursidae. Another 16 unplanned pregnancies (in Felidae, Ursidae, and Vivirridae) occurred because the institution failed to replace the MGA implant at the recommended 2-year interval.

Unplanned pregnancies also occurred as a result of lost implants. Ten percent of all MGA implants inserted into female carnivores were lost. In 12 percent of these cases, the lost implant led to an unintended pregnancy. A comparatively

Table 7.1
Contraceptive methods used, number of individuals, and number of treatments for carnivores[a]

Method	Number of delivered treatments	Number of ongoing treatments	Number of species	*n*
Steroid hormones:				
Progestin implants:				
MGA	2,589	242	59	1,557♀
Norplant II	4	2	4	4♀
Progestin injections:				
Depo-Provera	86	—	13	28♀
Proligestone	46	—	4	14♀
Oral progestins:				
Birth control pills[b]	3	0	3	3♀
Provera	8	3	3	6♀
Gestapuran	2 (1+)	2 (1+)	2	2♀+
MGA feed	1	1	1	1♀
Ovaban/Megace	34	0	9	21♀/1♂
Ovarid	6	0	2	2♀
Protein hormones:				
Deslorelin	15	0	3	9♀/5♂
Histrelin	3	0	2	4♀/2♂
Lupron	52	—	3	2♀/2♂
Immunocontraception:				
PZP vaccine	65	—	15	39♀
Mechanical:				
Vas plugs	6	0	4	6♂

MGA, melengestrol acetate; PZP, porcine zona pellucida.

[a]Data from the Contraception Database and PZP Database.

[b]Formulations: one progestin only and two unknown.

high percentage of MGA implants inserted into mustelids (n = 42) were lost (19 percent) or removed for a medical reason (22 percent; for example, abscess or infection at implant site). One species, the Oriental small-clawed otter (*Amblonyx cinereus*), accounted for 76 percent of the cases. Observations reported in the Contraception Database suggest that this could be a species-specific problem stemming from the otters' habit of rubbing against objects and the prolonged healing time resulting from the high percentage of time spent in water. Three other taxa also had relatively high loss rates: Ursidae (15 percent), Hyaenidae (19 percent), and Viverridae (19 percent). The masked palm civet (*Paguma larvata*) accounted for 80 percent of the lost implants within the viverrids.

Of the completed MGA implant treatments, 6 percent were stopped for a medical reason including a problem at the implant site (infection, abscess, skin reaction; 72 percent), uterine problems (pyometra, endometriosis; 19 percent), and mammary problems (hyperplasia, adenocarcinoma; 13 percent), or because females developed diabetes (1 percent).

Contraception with MGA implants was ended to allow reproduction in a total of 123 females in four carnivore families (Felidae, Canidae, Ursidae, and Mustelidae). Of these, 70 females (57 percent) conceived between 1 and 48 months after implant removal (Table 7.2). Most of the litters produced by these females (88 percent) were composed of live-born infants, 9 percent were composed of stillbirths, and in 2 percent of the litters the infant status was unknown or included live-born and stillborn infants. As the normal proportion of live births to stillbirths was not calculated for these taxa, the effect of MGA on fetal development was not evaluated. Of the 54 (45 percent) incomplete reversal attempts, 50 percent were ended for various reasons (for example, death, permanent sterilization, resumed birth control), and 50 percent are still considered ongoing reversal attempts.

Less than 1 percent of the carnivores (14 of 1,313) that received MGA implants were pregnant at the time of implant insertion. Of those 14 females, 12 were felids, 1 was a hyenid, and 1 was a canid (Table 7.3). Half the females (n = 8) were implanted with MGA during the first trimester of pregnancy, 3 during the second trimester, and 3 during the third trimester. Progestins are known to quiet myometrial contractions in some taxa (see Chapter 6, Choosing the Most Appropriate Contraceptive). However, with the possible exception of one African wild dog (*Lycaon pictus*) female, treatment with MGA for a portion or the duration of the pregnancy did not result in complications during parturition (see Table 7.3). In this single female, it was suspected that the MGA implant caused problems with parturition because the female failed to give birth around the expected delivery date. Five days after the implant was removed, the female had a stillbirth.

Table 7.2

Successful reversals for carnivores treated with contraceptives[a]

Method	Genus	*n*	Mean (range) duration of contraceptive use, months	Mean (range) time to conception, months[b]
Steroid hormones:				
Progestin implants:				
MGA	*Caracal*	2	18.75 (14.5–23)	1.5 (1.5)
	Felis	1	25	1
	Leptailurus	1	5	8
	Neofelis	2	28 (27[c]–28.5)	16 (13–19)
	Panthera	35	33.1 (5–141[c,d])	9.7 (<0.5–48[d])
	Puma	5	34.5 (12–67)	4.5 (<1–11)
	Uncia	14	26.7 (7.5–49.5)	14.9 (3–36)
	Canis	2	23 (16–30)	12 (9–15)
	Chrysocyon	3	13.7 (5–26[c])	16 (5–32)
	Lycaon	2	6	6 (5–7)
	Helarctos	1	18	5
	Tremarctos	1	20	9
	Ursus	1	33	19
Progestin injections:				
Depo-Provera	*Panthera*	2	6.25 (6–6.5[c,e])	3 (1.5–4.5[e])
Oral progestins:				
Birth control pills	*Leopardus*[f]	1	7.5	3
	Panthera	1	50	0.5
	Puma[g]	1	8.5	12
Provera	*Panthera*	1	27	8.5
Immunocontraception:				
PZP	*Panthera*	1	25	24
	Puma	1	34	<1
	Canis	2	30.25 (13.5–47)	27 (4–50)

MGA, melengestrol acetate; PZP, porcine zona pellucida.

[a]Data from the Contraception Database and PZP Database.

[b]Conception data calculated using gestation lengths from Hayssen et al. (1993).

[c]Indicates more than one consecutive contraceptive bout.

[d]One female's contraceptive bout was not ended to allow reproduction, but she conceived after birth control stopped and before start of new bout.

[e]Both contraceptive bouts not ended to allow reproduction, but both females conceived after birth control stopped and before start of new bout.

[f]Progestin-only birth control pills.

[g]Unknown birth control formulation.

Table 7.3
Successful parturition in carnivores implanted with MGA during pregnancy[a]

Genus/species	*n*	Trimester contraception started	Trimester contraception stopped	Successful parturition (*n*)
Steroid hormones:				
Progestin implants:				
MGA implant:				
Panthera leo	5	1	—	5[b]
	1	2	—	1
	1	3[c]	—	1
Panthera tigris	1	1	—	1
	1	2	—	1
	2	3	—	2
Uncia uncia	1	2	3	1
Lycaon pictus	1	1	3	1[d]
Crocuta crocuta	1	1	—	1

MGA, melengestrol acetate.

[a]Data from the Contraception Database.

[b]Breech birth.

[c]Female gave birth around insertion date; she was with male before implant insertion.

[d]Suspected that insertion of implant immediately following estrus resulted in delayed parturition.

Table 7.1 shows that only 9 percent of the 2,589 MGA implants used in carnivores represent ongoing bouts, in contrast with primates (see Chapter 8, Contraception in Nonhuman Primates) in which 25 percent of the total MGA implants used were ongoing bouts. Despite the high efficacy rates of MGA implants (and other progestins) in carnivores, these figures reflect a significant decrease in MGA implant use in this taxon. The reason for the decline is that research has shown progestin treatment can increase the risk of uterine and mammary pathology in felids and canids (Baldwin et al. 1994; Boer et al. 1996; Harrenstein et al. 1996; Kollias et al. 1984; Kazensky et al. 1998; Munson et al. 2002; see Chapter 5, Adverse Effects of Contraceptives). Munson et al. (2002) compared the reproductive tracts of 99 MGA-treated and 113 control (no exposure to contraceptives) female felids housed at 78 North American zoos. The authors showed that MGA treatment led to earlier occurrence of endometrial hyperplasia and more serious uterine lesions.

Additional problems with steroidal contraceptive use in carnivores are evident from canid studies that have shown there is a negative synergistic effect when estrogen is combined with progestin treatment (Giles et al. 1978; Teunissen 1952). Because of this amplified effect of estrogen on progestins, treatment of female canids with synthetic progestin contraceptives should never be initiated during the proestrous or estrous periods when endogenous estrogens are elevated (Brodey and Fidler 1966; Asa 1995). Kazensky et al. (1998) suggested that this confounding effect of estrogens on progestins could also be occurring in female felids that are treated with insufficient levels of MGA. The authors found no differences in the number of tertiary follicles in contracepted and noncontracepted female felids, indicating that folliculogenesis was not suppressed and ovulation was occurring at the tested MGA doses.

Although these health risks are most commonly reported for canids and felids, the same may be true for all carnivores. For example, Chittick et al. (2001) reported pyometra and adenocarcinoma in a female coati (*Nasua nasua*) that was treated with an MGA implant for 4.5 years. In addition, progestins have been linked with stimulation of insulin production (Frank et al. 1979) and diabetes (Nelson and Kelly 1976).

PROGESTINS: OTHER IMPLANTS The Contraception Database contains information on only one other progestin implant, the Norplant II implant (Wyeth-Ayerst; two-rod system with each rod containing 70 mg levonorgestrel; see Table 7.1). There were no failures in the four carnivores, all felids, administered a Norplant II implant, but the sample size is too small for a meaningful evaluation.

Looper et al. (2001) examined fecal estradiol levels in both treated and control domestic cats to determine whether levonorgestrel (LNG) would effectively eliminate follicular activity and subsequent mating and luteal activity in felids. They found that LNG given as a cesium-irradiated, slow-release injectable matrix was effective at preventing endogenous increases in estradiol. However, health risks with this progestin were identified by Baldwin et al. (1994), who reported that LNG implants in female domestic cats housed with males did not negatively affect body weight, physical mammary gland structure, or serum glucose levels but did cause uterine pathology including glandular epithelial hypertrophy and hyperplasia, endometrial cysts, and pyometra in most of the treated individuals.

PROGESTINS: INJECTIONS The Contraception Database contains information on two progestins administered by injection that have been used in carnivores, Depo-Provera (MPA) and proligestone, a second-generation progestin not produced in the United States (see Table 7.1). Evaluation of both products is difficult because, in too many instances, respondents to the Contraception Database questionnaire did not provide the dose (mg/kg body weight of the treated animal) or treatment interval. When such information was provided, the small sample size, the variation in dose, and number of single-use cases (yielding no data on interval) preclude generalizations.

A total of 28 females of 13 species were treated with Depo-Provera. Most of the 86 Depo-Provera treatments were administered to Ursidae (54 percent), followed by Felidae (38 percent), Canidae (5 percent), Mustelidae (2 percent), and Viverridae (1 percent). Doses ranged from 1.4 to 7.5 mg/kg body weight; no failures were reported. Depo-Provera treatment was stopped to allow reproduction in 2 lions, and both successfully conceived (see Table 7.2). Proligestone was only reported used in 4 Felidae species including tiger, jaguar, Eurasian lynx (*Lynx lynx*), and jaguarundi (*Herpailurus yagouaroundi*). Of the 46 treatments, the dosage used ranged from 7.4 to 42 mg/kg body weight; there were no failures.

PROGESTINS: ORAL Six orally delivered progestin contraceptive methods have been used in carnivores: medroxyprogesterone acetate (Provera: Pharmacia and Upjohn) used in two felid species; Gestapuran (LEO Pharma) used in two felid species; MGA in feed (Purina Mills) used in a sun bear (*Helarctos malayanus*); megestrol acetate (Ovaban: Schering-Plough; and Megace: Bristol-Myers Squibb), used in five felid, two canid, and two ursid species; megestrol acetate (Ovarid: Schering-Plough) used in a jaguar and a brown bear (*Ursus arctos*); and three types of human birth control pills (Clinovir, and two brands not identified) used in three felid species. In no case was the sample size per species large enough or the duration of use long enough to allow for meaningful evaluation (see Table 7.1). Four of six females (all felids) successfully reversed after termination of oral progestin contraception (see Table 7.2).

Protein Hormones

Three GnRH agonists have been used in zoo-housed carnivores. Research with one product, the deslorelin implant (Suprelorin: Peptech Animal Health) was carried out in domestic dogs and cats before its use in exotic carnivores. Trigg et al. (2001) demonstrated that long-term use of sustained-release deslorelin implants effectively and reversibly suppressed reproductive function in female domestic dogs. The duration of efficacy in females was not found to be dependent on the stage of the estrous cycle at the time of treatment initiation. Research trials with female domestic cats (Munson et al. 2001) showed that deslorelin implants were effective at suppressing estrous cycles, but individual differences in the duration of suppression were found.

Based on the positive results with domestic carnivores, Bertschinger et al. (2001, 2002) initiated a study to examine the efficacy of the deslorelin implant in exotic female carnivores including cheetahs (*Acinonyx jubatus*, n = 13), lions (n = 10), leopards (n = 3), and African wild dogs (*Lycaon pictus*, n = 15) housed in African breeding centers, and in gray wolves (*Canis lupus*, n = 5), red wolves (*C. rufus*, n = 5), and fennec fox (*Vulpes zerda*, n = 7) housed in North American zoos and wildlife sanctuaries (note Table 7.1 includes 5 of the fennec fox and 2 red wolf females used in this study). Deslorelin was an effective contraceptive in all the felids. A small percentage of the treated females experienced a brief estrus following implant insertion (see Chapter 3, Types of Contraception) but were suppressed thereafter for up to 21 months (Bertschinger et al. 2002). Among the canids, deslorelin was administered too close to the initiation of the wolf's breeding season to suppress ovulation, whereas all the fennec fox females were effectively suppressed. Two African wild dogs in proestrus and 1 in anestrus ovulated after being implanted with deslorelin, and 1 female conceived. Duration of efficacy (based on signs of estrus) varied between 3 and 21 months after treatment in 13 of the females.

Limited information was available on the use of histrelin, a GnRH agonist (The Population Council) that is no longer available (red wolf and clouded leopard, *Neofelis nebulosa*) and leuprolide (Lupron Depot: TAP Pharmaceuticals; lion and clouded leopard) as contraceptives in zoo carnivores (see Table 7.1). No failures were reported.

IMMUNOCONTRACEPTION

The only immunocontraceptive method used in zoo carnivores is the porcine zona pellucida (PZP) vaccine (see Table 7.1). In this summary, the initial series of injections, typically two or three, was counted as a single treatment because effective contraception is not achieved until completion of this series. Each follow-up booster injection (that is, at an 8-month or a 1-year interval) was counted as a single treatment. Of the 65 treatments reported administered in the PZP Database, 35 percent were the initial booster series and 46 percent were follow-up booster injections.

Of the 65 PZP treatments, 20 were performed in Felidae, 3 in Canidae, 14 in Ursidae, 1 in Procyonidae, and 1 in Mustelidae. Within Felidae, use of the PZP vaccine was discontinued because the females continued to engage in sexual activity, which resulted in copulatory-induced ovulation and pseudopregnancies (J. Kirkpatrick, personal communication). Those pseudopregnancies, associated with high progesterone concentrations, present as serious a health risk as do the contraceptives based on synthetic progestins. Thus, in carnivores with induced ovulation, PZP contraception is not a significant improvement over the progestins. The data within Canidae are inconsistent and very limited. The majority of Ursidae in the PZP Database were only recently contracepted, so data are incomplete. Within Procyonidae and Mustelidae, only single females are represented, which is insufficient for evaluation of the contraceptive. In general, the number of repetitions per species and per family is low, making it impossible to fully evaluate this contraceptive.

Three failures occurred with PZP use in carnivores; 2 felids and 1 canid became pregnant. One unintended birth, in a felid, resulted from the failure of the institution to administer the contraceptive at the recommended interval. Three additional unintended births occurred, in 2 ursids and 1 felid, but it was not possible to conclusively place them in either of the prior categories. Of the 39 females treated with the PZP vaccine, it is unclear how many were placed in a breeding situation following PZP treatment. The PZP Database does show that 4 females, 2 felids and 2 canids, successfully reversed once treatments were stopped (see Table 7.2).

Testing of contraceptive vaccines in female domestic dogs directed at the zona pellucida has revealed undesirable generalized ovarian effects that may cause not only suppression of cycles but also infertility (Mahi-Brown et al. 1982, 1985; Gonzalez et al. 1989), suggesting caution in regard to its use in canids.

METHODS FOR MALES

Carnivore contraception has primarily been directed at females, with castration and sterilization by vasectomy as the only practical methods available for males. However, vasectomy is not a good alternative for felids because of the health risks it can cause for female partners (see Chapter 6, Choosing the Most Appropriate Contraceptive; AZA CAG 2004).

Steroid Hormones

PROGESTINS The Contraception Database contains only one record of a male carnivore being treated with a progestin. A male Canadian lynx (*Lynx canadensis*) was treated with Ovaban; however, treatment was stopped after 1 month because of his lethargy and anorexia.

Protein Hormones

The Contraception Database reflects the paucity of male contraceptive alternatives currently available for carnivores, but this is likely to change because promising research results have been achieved with GnRH agonists. Trigg et al. (2001) showed that deslorelin suppressed spermatogenesis in domestic dogs in a dose-related manner. The treatment was considered to be reversed in all 16 treated males based on plasma testosterone, testis measurements, semen analysis (6 males evaluated), and fertile matings (2 males evaluated). These positive results led to research with deslorelin in exotic male carnivores, including 2 bush dogs (*Speothos venaticus*), 3 red wolves (see Table 7.1), and 3 sea otters (*Enhydra lutris*), all housed in zoos. Data from additional trials carried out with exotic carnivores, reported in Bertschinger et al. (2002), included red wolves ($n = 7$), gray wolves ($n = 5$), a black-footed cat (*Felis nigripes*, $n = 1$), African wild dogs ($n = 6$), cheetahs ($n = 6$), and a leopard ($n = 1$). Contraception was achieved in male wild dogs and cheetahs, and, although fertility was not examined, testosterone levels and testis size were reduced in the sea otter and the black-footed cat. As was true for female gray wolves, male wolves implanted at the onset of the breeding season were not suppressed; this was likely because the initial period of stimulation, characteristic of agonists, was almost as long as the very brief breeding season (about 6 weeks), in addition to the time needed for sperm in the vas to be cleared. Initiation of treatment well in advance of the breeding season, before females are fertile, should address this problem. Reproductive parameters were not suppressed in the 2 bush dogs at either the 3- or 6-mg dose. Although varied, these results are nevertheless encouraging overall and call for continued research on the application of this method with males. As indicated in Table 7.1, two other GnRH agonists, Lupron Depot and histrelin (The Population Council; no longer available) have been tested in captive carnivores. Although the sample size was very small ($n = 2$ per method), no failures were reported.

Mechanical Methods

Injectable vas deferens plugs, designed to block sperm passage through the vas deferens, were tested in four felid species (caracal, *Caracal caracal*; bobcat, *Lynx rufus*; snow leopard, *Uncia uncia*; and lion) and one ursid species (polar bear, *Ursus maritimus*) (see Table 7.1). The difficulties encountered with inserting the plugs, which caused irreversible damage to the vas deferens, ended the trials (L. Zaneveld and C. Asa, personal communication).

SUMMARY

The need for managing reproduction in zoo felids led to the development of the MGA implant, which today is the most widely used form of contraception in carnivores as well as in other species in US zoos. Indeed, MGA implants are highly effective in carnivores, as are the other progestin contraceptives if properly administered. Nevertheless, an alternative to progestin contraception is greatly needed because health risks are associated with the administration of sex steroids in carnivores (see Chapter 5, Adverse Effects of Contraceptives). For example, Munson et al. (2002) found that uterine pathology in exotic felids was significantly higher when MGA implant use exceeded 72 months. For this reason, the AZA CAG (2004) has recommended that, in those situations requiring contraception, treatment with MGA or other progestins not exceed 4 years. Preferably, when contraception is required, a management plan could be developed that limits each female's exposure to progestins to a period of 2 years, followed by a pregnancy before resuming progestin treatment.

PZP may not be an appropriate alternative as it failed to suppress estrus in most of the treated felids, although data for other carnivore species were limited. Because reduced follicular development, such as was seen in domestic dogs (Mahi-Brown et al. 1982), could lead to infertility, this method may not be a good contraceptive for carnivores.

Currently, the most promising new contraceptive method for use in carnivores, considering both efficacy and safety, is the GnRH agonist deslorelin. Preliminary trials are encouraging, and more research with deslorelin should be directed at determining effective dose and duration of efficacy for both males and females of different species.

ACKNOWLEDGMENTS

This summary would not have been possible without the time and effort invested by all the institutions that participated in the research program and completed annual CAG surveys. The technical assistance of Bess Frank in some of the preliminary Contraception Database summaries was greatly appreciated. The comments provided by anonymous reviewers on earlier drafts of this chapter were valued.

REFERENCES

American Zoo and Aquarium Association Contraception Advisory Group (AZA CAG) Annual Recommendations. 2004. www.stlzoo.org/contraception

Asa, C. S. 1995. Physiological and social aspects of reproduction of the wolf and their implications for contraception. In *Ecology and Conservation of Wolves in a Changing World. Part V. Canada: Behavior and Social Interaction*, ed. L. N. Carbyn, S. H. Fritts, and D. R. Seip, 283–286. Edmonton, Alberta, Canada: Canadian Circumpolar Institute, University of Alberta.

Baldwin, C. J., A. T. Peter, W. T. K. Bosu, and R. R. Dubielzig. 1994. The contraceptive effects of levonorgestrel in the domestic cat. *Lab Anim Sci* 44 (3): 261–269.

Bertschinger, H. J., C. S. Asa, P. P. Calle, J. A. Long, K. Bauman, K. DeMatteo, W. Jöchle, T. E. Trigg, and A. Human. 2001. Control of reproduction and sex related behaviour in exotic wild carnivores with the GnRH analogue deslorelin: preliminary observations. *J Reprod Fertil Suppl* 57:275–283.

Bertschinger, H. J., T. E. Trigg, W. Jöchle, and A. Human. 2002. Induction of contraception in some African wild carnivores by downregulation of LH and FSH secretion using the GnRH analogue deslorelin. *Reproduction Suppl* 60:41–52.

Boer, M., K. Cahnen, and M. Burkhard. 1996. A review of hitherto applied methods of contraception in zoological gardens. *Zool Gart* 66:93–105.

Brodey, R. S., and I. J. Fidler. 1966. Clinical and pathological findings in bitches treated with progestational compounds. *J Am Vet Med Assoc* 186:783–788.

Chittick, E., D. Rotstein, T. Brown, and B. Wolfe. 2001. Pyometra and uterine adenocarcinoma in a melengestrol acetate-implanted captive coati (*Nasua nasua*). *J Zoo Wildl Med* 32:245–251.

Frank, D. W., K. T. Kirton, and V. R. Berliner. 1973. The extrapolation of experimental findings (animal to man). The dilemma of the systematically administered contraceptive. *Lab Invest* 28:383.

Giles, R. C., R. P. Kwapien, R. G. Geil, and H. W. Casey. 1978. Mammary nodules in beagle dogs administered investigational oral contraceptive steroids. *J Natl Cancer Inst* 60:1351–1364.

Gonzalez, A., A. F. Allen, K. Post, R. J. Mapletoft, and B. D. Murphy. 1989. Immunological approaches in dogs. *J Reprod Fertil Suppl* 39:189–198.

Harrenstien, L. A., L. Munson, U. S. Seal, and American Zoo and Aquarium Association Mammary Cancer Study. 1996. Mammary cancer in captive wild felids and risk factors for its development: a retrospective study of the clinical behavior of 31 cases. *J Zoo Wildl Med* 27:468–476.

Hayssen, V., A. van Tienhoven, and A. van Tienhoven. 1993. *Asdell's Patterns of Mammalian Reproduction: A Compendium of Species-Specific Data*. Ithaca: Comstock.

Kazensky, C. A., L. Munson, and U. S. Seal. 1998. The effects of melengestrol acetate on the ovaries of captive wild felids. *J Zoo Wildl Med* 29:1–5.

Kollias, G. V. Jr., M. B. Calderwood-Mays, and B. G. Short. 1984. Diabetes mellitus and abdominal adenocarcinoma in a jaguar receiving megestrol acetate. *J Am Vet Med Assoc* 185 (11): 1383–1386.

Looper, S., G. Anderson, Y. Sun, A. Shukla, and B. Lasley. 2001. Efficacy of levonorgestrel when administered as an irradiated, slow-release injectable matrix for feline contraception. *Zoo Biol* 20:407–421.

Mahi-Brown, C. A., T. T. F. Huang Jr., and R. Yanagimachi. 1982. Infertility in bitches induced by active immunization with porcine zonae pellucidae. *J Exp Zool* 222:89–95.

Mahi-Brown, C. A., R. Yanagimachi, J. C. Hoffman, and T. T. F. Huang Jr. 1985. Fertility control in the bitch by active immunization with porcine zonae pellucidae: use of different adjuvants and patterns of estradiol and progesterone level in estrous cycles. *Biol Reprod* 32:761–772.

Munson, L., J. E. Bauman, C. S. Asa, W. Jöchle, and T. E. Trigg. 2001. Efficacy of the GnRH analogue deslorelin for suppression of oestrous cycles in cats. *J Reprod Fertil Suppl* 57:269–273.

Munson, L., I. A. Gardner, R. J. Mason, L. M. Chassy, and U. S. Seal. 2002. Endometrial hyperplasia and mineralization in zoo felids treated with melengestrol acetate contraceptives. *Vet Pathol* 39:419–427.

Nelson, L. W., and W. A. Kelly. 1976. Progestogen-related gross and microscopic changes in female beagles. *Vet Pathol* 13:143–156.

Seal, U. S., R. Barton, L. Mather, K. Olberding, E. D. Plotka, and C. W. Gray. 1976. Hormonal contraception in captive female lions (*Panthera leo*). *J Zoo Anim Med* 7:12–20.

Teunissen, G. H. B. 1952. The development of endometritis in the dog and the effect of oestradiol and preogesterone on the uterus. *Acta Endocrinol* 9:407–420.

Trigg, T. E., P. J. Wright, A. F. Armour, P. E. Williamson, A. Junaidi, G. B. Martin, A. G. Doyle, and J. Walsh. 2001. Use of a GnRH analogue implant to produce reversible long-term suppression of reproductive function in male and female domestic dogs. *J Reprod Fertil Suppl* 57:255–261.

INGRID J. PORTON AND KAREN E. DEMATTEO

8
CONTRACEPTION IN NONHUMAN PRIMATES

The systematic use of reversible contraception as a management tool for primates housed in zoological parks began in the mid-1980s. Some of the first species to be contracepted were those managed through a Species Survival Plan (SSP), such as the golden lion tamarin (*Leontopithecus rosalia*), lion-tailed macaque (*Macaca silenus*), orangutan (*Pongo pygmaeus*), and ruffed lemur (*Varecia variegata*). Today all primate species housed in American Zoo and Aquarium Association (AZA)-accredited zoos are managed at some level through a Taxon Advisory Group's Regional Collection Plan (RCP). Under an RCP, species not selected to be managed under an SSP or a Population Management Plan (PMP) are nevertheless assigned to a management category such as Research or Phase Out.

Because space is the greatest limiting factor to accomplishing RCP goals, fertility control remains an important management issue for all primate taxa. As all primates, even those that exhibit more solitary social or mating systems in the wild, live within an enriching social network, isolating individuals to prevent reproduction can be contrary to their basic social needs. In fact, most primates are found in complex social groups, and the continual separation of individuals from the group to achieve contraception can be very disrupting and even dangerous to the targeted individuals, other group members, or both. Contraception also permits managers to retain young primates in their natal group past sexual maturity without the negative consequences of inbreeding. Reversible contraception has become an important component of responsibly and humanely managed captive breeding programs for primates.

This chapter reviews the current methods used to contracept nonhuman primates and incorporates a summary of the Contraception Advisory Group (CAG)

Contraception Database as of December 2001. Definitions and methods used to collect the data are described in the introduction to Part III.

The Contraception Database contains information on 87 primate species including representatives from the four major primate groups: prosimian, New World monkey, Old World monkey, and ape. Although the results are presented separately for these four taxonomic groups, the data for New World monkeys are frequently subdivided by family (Cebidae and Callitrichidae) to provide greater interpretive clarity. The number of individuals (sample size) per taxon varies from 1 (17 taxa) to 215 (ruffed lemur), with only 18 taxa that have more than 50 contracepted individuals (see Appendix). Forty methods of reversible contraception have been used with primates, but most are steroidal hormones, primarily progestins (Table 8.1). The methods include 4 types of progestin implants, 1 form of injectable progestin, 6 progestin-only pills, 2 estrogen–progestin implants, 21 combination estrogen–progestin birth control pills, 3 gonadotropin-releasing hormone (GnRH) agonists, 1 immunocontraceptive method, and 2 mechanical methods. Of these, the melengestrol acetate (MGA) implant (supplied by E. D. Plotka) is the method most commonly used, with two exceptions; Depo-Provera is more frequently used in prosimians, and the human birth control pill is used preferentially in two ape species, chimpanzees (*Pan troglodytes*) and gorillas (*Gorilla gorilla*). For the majority of primate species that, in captivity, are capable of reproduction throughout the year, the data presented in this chapter are summarized and discussed as number of months. For seasonal species, such as lemurs, efficacy and reversibility are more appropriately reported as number of seasons.

METHODS FOR FEMALES

Steroid Hormones

PROGESTINS: MELENGESTROL ACETATE IMPLANTS The Contraception Database contains records for 3,189 MGA implants that have been inserted into nonhuman primates (see Table 8.1). Old World monkeys (40 percent) and New World monkeys (34 percent) accounted for the majority of MGA implants used in primates. Within New World monkeys, 77 percent of MGA implants were used in callitrichids. Ape species accounted for 13 percent and prosimians for 12 percent of MGA implant use. Of this total, 774 MGA implant bouts are considered ongoing, meaning that at the time the data were reported the implants were believed to be in place and effective. Another 506 implants were lost at some time after insertion.

Table 8.1
Contraceptive methods, number of individuals, and number of treatments for primates[a]

Method	Number of delivered treatments	Number of ongoing treatments	Number of species	n
Steroid hormones:				
Progestins:				
MGA Implant	3,007	755	84	1,843♀/2♂
Norplant	113	53	17	110♀
Implanon	37	30	5	32♀
Jadelle	6	5	1	6♀
Depo-Provera	2,505	—	49	397♀
Birth control pills[b]	13	0	3	12♀
Ovaban/Megace	9	4	4	8♀
Steroid hormones:				
Estrogen–progestin combination				
MGA–estrogen implant	21	0	4	19♀
Birth control pills	161	84	6	145♀
Protein hormones:				
GnRH agonists:				
Deslorelin	19	7	5	9♀/10♂
Histrelin	4	—	1	4♂
Lupron Depot	12	—	3	1♀/2♂
Immunocontraception:				
PZP	94	0	10	22♀
Mechanical:				
IUD	1	0	1	1♀
Vas plugs (♂)	15	0	11	15♂

MGA, melengestrol acetate; GnRH, gonadotropin-releasing hormone; PZP, porcine zona pellucida; IUD, intrauterine device.

[a]From Contraception Database.

[b]Progestin-only pills.

Data for completed bouts (implant inserted and removed) are available for 1,909 MGA implants. The recommended duration of MGA implant use is 2 years (AZA CAG 2004). Twenty-five percent of the implants were removed at 12 months or earlier for a variety of reasons (for example, animal transfer, contraception no longer required, change of method, irritation or abscess at implant site, death). The majority of MGA implants placed in primates (40 percent) were used for 13 to 24 months, 21 percent remained in use for 24 to 31 months, 5 percent were removed between 32 and 36 months, and 9 percent were left in place for more than 3 years. For seasonally breeding species such as lemurs, the CAG has recommended MGA implants be used for a single breeding season as opposed to keeping the implants in place throughout the entire year (AZA CAG 2004). The prosimian data reflect this, with 36 percent of the implants in use for less than 1 year.

Fifty-six MGA implant failures, defined as the implant confirmed to be in place at the time of conception, were identified. Of these, more than half (57 percent) of the failures were in Cebidae. When the data are presented as the percentage of failures compared to the completed MGA implant bouts per primate group, the degree of the problem with cebid species is more easily recognized. Thus, the failure rate was 0.4 percent in prosimians, 1.1 percent in Old World monkeys, and 2.6 percent in apes. Among New World monkeys, the failure rate of 1.3 percent in callitrichids was fairly equivalent to other taxa, in contrast to the 18.4 percent failure rate in cebids. These latter results are consistent with those from an earlier review of the Contraception Database that identified a disproportionate rate of MGA implant failures in Cebidae (Porton et al. 1990; Porton 1995; Asa et al. 1996). New World monkeys have higher levels of endogenous sex steroids than Old World monkeys (Chrousos et al. 1984; Coe et al. 1992), but because MGA implant doses were originally based on the Old World monkey model, the failures are believed to be dose related. This conclusion has been supported with a more detailed analysis of the Contraception Database that found small-bodied callitrichids, which were receiving relatively higher MGA doses, had correspondingly lower failure rates (I. Porton and K. DeMatteo, unpublished data). A study monitoring serum progesterone levels determined that ovarian cyclicity was completely suppressed in five adult female cotton-top tamarins (*Saguinus oedipus*) treated with MGA implants (Lee-Parritz and Roberts 1994). Similar conclusions were reached when the efficacy of MGA implants was assessed through urinary progestin levels in two other callitrichid species, the common marmoset (*Callithrix jacchus*; Moehle et al. 1999) and the golden-headed lion tamarin (*Leontopithecus chrysomelas*; Van Elsacker et al. 1994). As a result of the evidence that a higher dose of MGA is effective at preventing pregnancy in one group of New

World monkeys, callitrichids, the MGA implant dose for Cebidae was increased in the late 1990s. Although no pregnancies have been reported, the sample size is small and complete evaluation awaits additional data.

The current and previous (Porton et al. 1990) summaries of the Contraception Database have shown that the most significant problem with MGA implants is the loss rate. To reduce MGA implant loss, the CAG recommendations state that primates should be separated for 5 to 7 days following implant insertion to prevent allogrooming and implant removal by conspecifics (AZA CAG 2004). In the case of small-bodied primates, steel sutures that prevent implant removal can preclude the need to separate individuals. The current review of MGA implant loss, defined as the percentage of implants lost compared to the total number inserted per primate taxon, showed that prosimians (28 percent) had the highest implant losses, followed by apes (21 percent), Old World monkeys (16 percent), and New World monkeys (9 percent). In prosimians, the loss problem was greatest in brown lemurs (*Eulemur fulvus*, 54 percent) and ringtailed lemurs (*Lemur catta*, 42 percent). Within apes, orangutans (30 percent) and chimpanzees (29 percent) had the highest losses, whereas the problem was much less in gibbons (*Hylobates* spp., 8 percent). In Old World monkeys, the loss problem was greatest in macaques, for example, 54 percent loss rate in rhesus macaques (*Macaca mulatta*) and 27 percent in lion-tailed macaques. In contrast, loss rates were much lower in Old World monkey leaf-eating species (Colobinae, 6 percent). Among the New World monkey species, the loss rate was highest in cotton-top tamarins (17 percent). The reason for implant loss is typically unknown, often because caretakers discover the implant missing some time after the fact. Thirty-three implants were reported as surgically removed because of an abscess at the implant site; it can be assumed that at least some implant losses were the result of unobserved infections. Orangutans accounted for 11 of the 33 (33 percent) reported abscesses, more than any other species, and implant loss rate is also high in this species.

Because it can be difficult for managers to know when an implant has been lost, the result may be unwanted births. A total of 90 primate births occurred due to lost implants. Summarized by group, 6 percent of implant losses led to unplanned births in apes, 16 percent in prosimians, 19 percent in Old World monkeys, 26 percent in callitrichids (all were cotton-top tamarins), and 47 percent in cebids.

The Contraception Database shows that MGA implants were removed from 180 females to allow reproduction; 111 of these females (62%) conceived after implant removal. Specifically, 25 of the 29 prosimians (86 percent), 36 of the 65 callitrichids (55 percent), 7 of 10 cebids (70 percent), 22 of the 42 Old World monkeys (52 percent), 7 of the 14 lesser apes (50 percent), and 14 of 20 great apes (70 percent; all attempts were with orangutans) became pregnant following MGA implant removal. The range and average time from MGA implant removal to conception for each genus is presented in Table 8.2. A more detailed study of the ruffed lemur SSP population found that 12 nulliparous and 10 parous females successfully conceived following one to six breeding seasons of contraception with MGA implants or Depo-Provera (Porton, in preparation).

Moehle et al. (1999) investigated MGA implant efficacy and reversibility through urinary progestin levels and the occurrence of pregnancy in 10 common marmosets. Females were treated and successfully contracepted for 6 to 8 or 19 to 21 months. Implant removal resulted in ovulation within 3 weeks and conception within 4 months in all females. De Vleechouwer et al. (2000a) evaluated the efficacy and reversibility of MGA implants in golden-headed lion tamarins. To examine reversibility, the authors included 5 females whose MGA implants were removed and another 19 females whose MGA implants they defined as expired because, although not removed, they were more than 2 years old, and the females were in breeding situations. Two females from which the MGA implants were removed conceived at 3 and 5 months following implant removal. Three females with MGA implants in place conceived at 25.5, 28, and 29 months after implant insertion. Of the 16 MGA-implanted females that did not conceive, time from presumed expiration (24 months) to the end of the study was 8 to 29 months.

In an extensive review of the reversibility of MGA implants in golden lion tamarins, Wood et al. (2001) found that reproduction was four times greater in females whose implants were removed versus those that had expired. Comparing the probability of breeding between formerly implanted versus nonimplanted females, the authors showed that when female age is controlled for (younger females have a higher probability of breeding, and the control group included more young females than the implanted group), there was no significant difference in the probability of breeding between the formerly MGA-implanted females compared to the control group. This research highlights that return to fertility rates must be evaluated within the context of a species normal life table fecundity rates in captivity; that is, 100 percent female fecundity does not occur in any captive species and should not be expected following cessation of contraceptive use.

A total of 156 primates received an MGA implant while pregnant (Table 8.3). These implants were typically inserted without the realization that the female was already pregnant, with the exception of the callitrichids. Many of the callitrichid females were purposefully implanted while pregnant to prevent the postpartum estrus that occurs in these species. Inserting an MGA implant into a female after she has given birth but before she undergoes her postpartum estrus is impractical for numerous reasons including endangering the infants' lives, disturbing the mother–infant bonding process, and complications of an incision at the same

Table 8.2
Successful reversals for primates treated with contraceptives[a]

Method	Genus	*n*	Mean (range) duration of birth control use, months[b]	Mean (range) time to conception,[c] months[b]
Steroid hormones:				
Progestins:				
MGA implants	*Eulemur*	13	1.5 (1–4)	1.6 (1–3)
	Varecia	5	1.6 (1–3)	2.6 (1–4)
	Lemur	2	1.5 (1–2)	1 (1)
	Propithecus	1	1	1
	Callimico	2	14 (9–19)	9.75 (9.5–10)
	Callithrix	7	9.5 (6–21.5)	2.1 (1–4)
	Leontopithecus	16	18 (5–26.5)	7.4 (1–25)
	Saguinus	11	34.5 (2–69)	5.2 (1–11)
	Pithecia	1	3	2
	Alouatta	3	10.8 (1–20)	2 (1–4)
	Ateles	2	7.75 (4–11.5)	16.5 (3–30)
	Callicebus	1	3.5	12
	Cercocebus	2	29 (29)	11.5 (7–16)
	Macaca	7	52.3 (14–98)	11.5 (5.5–20)
	Mandrillus	2	20.7 (1.5–40)	19 (3.5–35)
	Papio	1	24	20
	Colobus	7	34.5 (4.5–94)	7.7 (4–14)
	Pygathrix	2	23 (22–24)	10.5 (4–17)
	Trachypithecus	1	61	2
	Hylobates	7	44.7 (22–62)	3.8 (1–10)
	Pongo	14	33.4 (5–68)	21.1 (2–67)
Norplant	*Pithecia*	1	3	2
	Pongo	4	29 (14–45)	22 (4–57)
Depo-Provera	*Hapalemur*	1	<1	<1
	Eulemur	5	<1 (<1–1)	1.2 (1–2)
	Varecia	11	2.4 (1–6)	1.4 (1–2)
	Alouatta	2	4.3 (3–6)	14 (1–30)
	Ateles	1	4	1
	Macaca	4	6 (6)	9.5 (9–11)
	Erythrocebus	1	16	<11[d]
	Hylobates	1	3	15
	Pan	2	5.5 (5–6)	5.0 (4–6)
MGA implant and Depo-Provera	*Eulemur*	2	2.5 (2–3)	1 (1)
	Varecia	2	2.5 (2–3)	1 (1)

Continued on next page

Table 8.2 continued

Method	Genus	*n*	Mean (range) duration of birth control use, months[b]	Mean (range) time to conception,[c] months[b]
Estrogen–progestin combinations:				
Birth control pills	*Pongo*	4	25 (2–49)	7.5 (1–18)
	Pan	5	17.5 (13–25)	2.6 (1–5)
	Gorilla	5	18.7 (3.5–46)	23.6 (4–44)

MGA, melengestrol acetate.

[a]Data from the Contraception Database.

[b]Seasons, not months, for lemurs.

[c]Conception data calculated using gestation lengths from Hayssen et al. (1993).

[d]Female was transferred to another facility 11 months after DepoProvera treatment was stopped; female was pregnant but was lost to follow-up.

location where infants are carried. Not unexpectedly, callitrichid species comprise 45 percent of the primates implanted during pregnancy, prosimians 3 percent, cebids 13 percent, Old World monkeys 24 percent, and apes 15 percent. Ninety-six (62 percent) of the MGA implants were inserted during the first trimester, 33 (21 percent) during the second trimester, and 27 (17 percent) during the third trimester. Of these pregnancies, 138 resulted in live births (88 percent) and 18 in an abortion or a stillbirth (12 percent). Of the primates that received MGA implants during the first trimester, 10 percent aborted or had stillbirths. Abortions or stillbirths occurred in 6 percent and 10 percent of the females implanted with MGA during the second and third trimester, respectively. These results undoubtedly underreport the percentage of abortions or miscarriages, especially during the first trimester, because these events can go undetected. There were no reports that females administered contraception during their pregnancy experienced difficulties during parturition.

De Vleeschouwer et al. (2000a) used data from the International Studbook for Golden-headed Lion Tamarins and a follow-up survey to evaluate the effect of MGA implant contraception on neonatal birth status. They found, in seven females for which there were sufficient comparable data, that there was no significant difference in the number of stillborn versus live offspring born to the females before or during MGA contraception. The authors did find that stillbirth rates were significantly higher in four of these females when breeding was resumed

Table 8.3
Successful parturition in primates treated with a contraceptive during pregnancy[a]

Family	n	Trimester contraception started	Trimester contraception stopped	Successful parturition (n)
Steroid hormones:				
Progestins:				
MGA implant				
Lemuridae	4	1	—	4[b]
Callitrichidae	43	1	—	43
	2	1	1	2
	13	2	—	13
	13	3	—	13
Cebidae	15	1	—	15
	1	1	2	1
	1	2	—	1
	4	3	—	4
Cercopithecidae	15	1	—	15
	4	1	3	4
	12	2	—	12
	1	2	3	1
	5	3	—	5
Hylobatidae	5	1	—	5[b]
	6	2	—	6
	4	3	—	4
Pongidae	3	1	—	3[c]
	3	1	2	3
	1	1	3	1
	1	3	—	1
Norplant				
Cercopithecidae	1	1	Unknown	1
	2	2	—	2
	1	3	—	1
Pongidae	1	1	3	1
	1	2	—	1
Jadelle				
Cercopithecidae	2	1	Unknown	2
Depo-Provera				
Lorisidae	2	1	—	2
Lemuridae	7	1	—	7

Continued on next page

Table 8.3 continued

Family	*n*	Trimester contraception started	Trimester contraception stopped	Successful parturition (*n*)
	1	1	3	1
	2	2	—	2
	1	2	3	1
Callitrichidae	2	1	3	2
Cebidae	1	1	—	1
	1	1	3	1
	1	1	2	1
	1	3	—	1
Cercopithecidae	1	1	—	1
Pongidae	3	1	—	3[d]
	1	2	—	1
Progestin-only pill				
Cercopithecidae	1	1	—	1
	1	1	2	1
Hylobatidae	1	1	—	1
Pongidae	1	1	1	1
	1	1	3	1
Estrogen–progestin combination:				
Hylobatidae	1	1	2	1
Pongidae	6	1	—	6
	1	1	2	1
	2	1	3	2
	1	2	3	1

MGA, melengestrol acetate.

[a]Data from the Contraception Database.

[b]One female lost implant at unknown time during pregnancy.

[c]One female was implanted with an MGA implant and a Norplant implant.

[d]One female was contracepted with both Depo-Provera and birth control pills.

after MGA implant contraception. However, in a larger study of golden lion tamarins, Wood et al. (2001) compared like-aged formerly MGA-implanted to non-implanted females and found that the stillbirth rate between the two groups was not significantly different.

The effect of MGA implant contraception on the expression of sexual behavior in primates has not been well studied. Responses to the Contraception Database Surveys include scattered reports of mating observed between an MGA-implanted female and her partner. However, the observations are not systematic, the status of the implant is not always known, and the distinction between mounting behavior and an actual copulation is rarely provided. One study of seven MGA implant-treated female and two intact male hamadryas baboons (*Papio hamadryas*) revealed that, although contraception did not eliminate all sexual activity, an immediate increase in sexual behavior followed MGA implant removal (Portugal and Asa 1995). De Vleeschouwer et al. (2000b) showed that in four pairs of golden-headed lion tamarins breeding behavior continued, but at lower frequencies, after the females were administered MGA implants. Both these studies suggest a low level of sexual behavior is not atypical but that increased rates may be an indication of implant loss.

Certain primate species including chimpanzees and some Old World monkeys (baboons, mangabeys, and some macaque and guenon species) exhibit a sexual swelling of the skin around the genitalia during estrus. The relationship between sexual skin swellings and sexual behavior has long been recognized (Michael and Zumpe 1971). Copulations typically peak during but are not necessarily restricted to the period of maximal swelling (Michael and Zumpe 1971; Richard 1985). Increase of the sexual swelling to its full size and subsequent decrease to baseline defines the female's estrous cycle. Ovulation occurs at full tumescence, and the swelling provides males a clear visual cue of this physiological event. Consequently, the effect of contraception on sexual swellings and sexual behavior is likely interconnected. Contraception with MGA implants typically caused chimpanzee sexual swellings to cease; however, individual variation has been reported, with some females exhibiting slight swellings (Contraception Database). The sexual swellings of seven hamadryas baboon females were completely suppressed after MGA implant insertion. Resumption of sexual swellings occurred within 9 days of implant removal and corresponded to a parallel increase in sexual behavior (Portugal and Asa 1995).

The behavioral consequences of contraception on reproductive suppression in callitrichid family groups have important management ramifications. Normally, callitrichid daughters are behaviorally or physiologically suppressed from cycling while housed with their mother (Abbott et al. 1993; Chaoui and Hasler-Gallusser

1999). Whether contraception of the breeding female will remove this suppressive effect, or otherwise change behavioral interactions among females in a family group, is information needed for the development of management plans. Lee-Parritz and Roberts (1994) reported changes in the relationships between cotton-top tamarin females in two unisex pairs. In one pair, fighting broke out following the insertion of an MGA implant in the dominant, cycling female, and in the second unisex pair the subordinate, suppressed female exhibited ovulatory cycles after the dominant female was treated with an MGA implant. Within four golden-headed lion tamarin families, treating the mothers with an MGA implant did not alter sexual relations between male and female family members, nor did inserting an MGA implant in the breeding female change affiliative or agonistic social interactions among family members in a uniform and predictable manner (De Vleeschouwer et al. 2000b).

Investigating the effect of progestin contraception on allogrooming, another social behavior, in stump-tailed macaques (Steklis et al. 1983), hamadryas baboons (Portugal and Asa 1995), and golden-headed lion tamarins (De Vleeschouwer et al. 2000b) revealed the rates remained equivalent or just slightly altered after contraception. Interestingly, as seen in humans, there is some indication from respondents to the Contraception Database Survey that contraceptives appear to affect the mood of nonhuman primate females in different and individual ways, for example, comments that certain females were more aggressive or "cranky" when treated with a progestin.

To date, few studies have shown a link between MGA implant treatment and serious health risks in nonhuman primates. Munson et al. (see Chapter 5, Adverse Effects of Contraceptives) found no correlation between pathology and contraceptive use in Old World monkeys and apes. Gresl et al. (2000) reported that, despite the predisposition for glucose intolerance and diabetes in orangutans, the results of an intravenous glucose intolerance test did not differ between females treated versus those not treated with MGA implants. Murnane et al. (1996) linked an exuberant, aggressive decidual response and endometritis to MGA use in Goeldi's monkey (n = 17) and squirrel monkeys (n = 3). Contraception of Goeldi's monkeys with MGA implants has been discontinued and is not recommended (Murnane et al. 1996; see Chapter 5, Adverse Effects of Contraceptives). Moehle et al. (1999) used ultrasonography to examine the effect of MGA implants on the uterus of 10 common marmosets. Although changes commensurate with uterine hypertrophy and decidualization were observed, the changes were not permanent and all 10 females conceived after implant removal. Munson (see Chapter 5, Adverse Effects of Contraceptives) reports the occurrence of endometrial hyperplasia in Lemuridae treated with MGA implants, but whether this adversely impacts fertility is not yet known. MGA-treated lemurs have conceived after treatment (see Table 8.2). Reversibility in ruffed lemurs, the species with the most data, appears good. In a separate review of the Ruffed Lemur International Studbook, Porton (in preparation) found that 22 ruffed lemur females have successfully conceived after multiple bouts of MGA or Depo-Provera contraception, including nulliparous females conceiving at 7 to 11 years of age.

PROGESTINS: OTHER IMPLANTS The Contraception Database lists four other progestin-only implants that have been used in primates: Norplant (six-rod system with each rod containing 36 mg levonorgestrel; Wyeth-Ayerst) and Norplant II (two-rod system with each rod containing 70 mg levonorgestrel; Wyeth-Ayerst), Jadelle (one rod containing 75 mg levonorgestrel; Wyeth-Ayerst), and Implanon (one rod containing 68 mg etonogestrel; Organon). The Contraception Database contains limited data for the latter two, making a complete evaluation of those methods premature (see Table 8.2). Jadelle was used in 8 female baboons; 5 bouts are ongoing, and another was terminated at 24 months when the female died of causes unrelated to the contraception. Two female baboons were pregnant while implanted with Jadelle, although 1 may have conceived just days after implant insertion (see Table 8.3). Both females gave birth to live young, but implant status at the time of parturition was not reported. Implanon has been used in five primate species, including 25 chimpanzees, 5 gorillas, 4 orangutans, 3 Debrazza monkeys (*Cercopithecus neglectus*), and 1 black and white colobus (*Colobus guereza*). Thirty of 38 Implanon bouts are ongoing, whereas 7 bouts, all in chimpanzees, were completed between 21 and 37 months after insertion with no failures. One Implanon inserted in a chimpanzee was reported lost, and the female became pregnant.

Norplant has been used in 1 prosimian, New World monkeys, Old World monkeys, and apes, although the number of species and contraceptive bouts is small compared to those for MGA implants. There is little information on the effective dose or duration for Norplant implant use in species other than humans. As no specific recommendations for implant dose or removal have been available to guide Norplant contraception in nonhuman primates, the duration of use and dose (number of rods) has varied. The Contraception Database contains records for 57 completed Norplant implant bouts, 4 in New World monkeys ($\bar{x}$, 13.2 months; range, 3 to 23 months), 17 in Old World monkeys ($\bar{x}$, 25.7 months; range, 8 to 45 months), and 26 in apes ($\bar{x}$, 39.2 months; range, 4 to 66 months). Another 52 bouts are ongoing: 1 ruffed lemur, 7 in New World monkeys, 19 in Old World monkeys, and 25 in apes.

A total of 15 Norplant implant failures have been reported in the Contraception Database, 6 (40 percent) in New World monkeys, 4 (19 percent) in Old World monkeys, and 5 (13 percent) in apes. Nine of the monkey species failures were between 8 and 10 months after Norplant insertion: 2 cotton-top tamarins (70-mg dose), 3 white-faced saki monkeys (*Pithecia pithecia*; 70-mg, 72-mg, and unknown dose), 1 Francois langur (*Trachypithecus francoisi*; 36-mg dose), 1 silvered langur (*Trachypithecus cristatus*; 70-mg dose), and 2 baboons (36-mg dose). Another white-faced saki monkey conceived 15 months after Norplant (72-mg) insertion. These 9 failures are suggestive of an inadequate dose.

The five Norplant implant ape failures occurred at 52 to 54 months, in one orangutan (140-mg dose) and in four chimpanzees administered the six-rod (216-mg) system. In chimpanzees, the presence of a sexual swelling or sexual behavior cannot be used as a definitive indication that the implant is lost or ineffective. Bettinger et al. (1997) found that Norplant implants did not suppress chimpanzee sexual swellings; however, the size of the tumescence varied among females from partial to full, and sexual behavior continued at normal rates (Bettinger et al. 1997). Although there are chimpanzees that have been successfully treated for 60 months with a Norplant implant (Bettinger and DeMatteo 2001), which is the normal duration of efficacy in women (Bardin and Sivin 1992), the Contraception Database results suggest there may be more variability in apes. Consequently, Norplant implant replacement at 50 months may be an appropriately conservative protocol.

Only five females (four orangutans and one white-faced saki monkey) have had Norplant implants removed for reproduction. The white-faced saki monkey (70 mg) had been implanted for 3 months and conceived 2 months following implant removal, giving birth to a live infant. All four orangutans, implanted from 14 to 72 months, conceived between 4 and 57 months following implant removal. Three females gave birth to live young; the fourth had a stillbirth. Seven females were implanted with Norplant implants when already pregnant. Six females gave birth to live infants (three baboons and three chimpanzees, but one infant was premature), and one female baboon gave birth to a stillborn infant.

PROGESTINS: INJECTIONS The Contraception Database contains records for 49 nonhuman primate species that have been contracepted with medroxyprogesterone acetate (MPA, Depo-Provera): 13 prosimian, 16 New World monkey, 16 Old World monkey, and 4 ape species (see Table 8.1). However, in most primate species (76 percent), 5 or fewer individuals were treated with Depo-Provera. Furthermore, 75 percent of the contraception periods have comprised only 1 to 2 injections, indicating that for most species this method is used as an interim form of contraception (such as before the acquisition of an MGA implant). The limited information on repeated injections in these taxa makes evaluation of the contraceptive method premature. The majority of the Depo-Provera data are from its use in seasonally breeding species of prosimians (13 species, 245 individuals, covering 658 breeding seasons and more than 2,000 injections).

The challenge with using Depo-Provera in nonhuman primates is determining the dose and interval frequency required for efficacy in the diversity of species treated with this method. The human dose is 150 mg every 90 days (based on an average female weight of 59 kg, the dose is 2.5 mg/kg). Apes have most often been treated at this dose and interval, whereas the Contraception Database shows that doses given to Old and New World monkeys vary considerably (in numerous cases, however, the dose was not reported). Old World monkeys have been treated with doses from as low as 1.6 mg/kg to as high as 37 and even 75 mg/kg. The small-bodied New World monkeys have most often been treated with doses in the range of 9 to 30 mg/kg, but some received as much as 97 mg/kg. Doses for the larger New World monkeys were between 5 and 15 mg/kg. In addition, the efficacy of Depo-Provera could not be assessed in a number of both Old and New World monkeys that were given an injection at the same time an MGA implant was first inserted or at the end of 2 years when an MGA implant was still in place. Prosimians have been treated with a range of doses (2.5, 5, and 10 mg/kg), but since the late 1990s, 5 mg/kg is the dose typically used and recommended (AZA CAG 2004). In contrast to the majority of prosimians, which are hand caught for Depo-Provera injections, for most large primates the injection is delivered by dart, which decreases the certainty that the full dose was received and hinders interpretation of minimum effective dose.

The Contraception Database also shows that a large range of Depo-Provera treatment intervals have been used in nonhuman primates, with the range varying not only between but also within species and individuals. Lemurs have received as few as 1 to as many as 6 or 7 Depo-Provera injections within a single breeding season; for example, some ringtailed lemurs have been treated monthly for 5 to 6 months. Sample size for Depo-Provera use in Old and New World monkeys is small, and analyses are again complicated by disparate injection intervals, ranging from 1 to 3 months. Additionally, because many individuals were only treated with 1 or 2 injections, there is a paucity of data from which to develop recommendations for appropriate treatment dose and interval. Among apes, the number of injections administered to a single individual range from 1 to 25, and only 1 of the 9 gibbon (11 percent), 5 of the 13 orangutan (38 percent), and 2 of the 19 chimpanzee (11 percent) females were treated with Depo-Provera in successive treatment periods that exceeded 1 year. One orangutan female was scheduled on an every-60-day injection schedule, but most of the apes received injections at intervals that ranged between 80 and 95 days.

Depo-Provera may or may not inhibit mating behavior in nonhuman primates. For example, Steklis et al. (1983) found that stump-tailed macaque (*Macaca arctoides*) females treated with Depo-Provera did not copulate for 68 days after injection when in a semifree-ranging environment. Five free-ranging Japanese macaques treated with 19.4 mg/kg Depo-Provera did not exhibit copulatory behavior throughout the entire breeding season. In contrast, Depo-Provera-treated females housed in laboratory cages did copulate with males but at a lower frequency than untreated females. In a similar situation, forced proximity in laboratory cages was one explanation for the breeding behavior observed between newly paired ovariectomized stump-tailed macaque females and intact males (Goldfoot et al. 1978).

Twenty-one Depo-Provera failures at doses of both 5 and 10 mg/kg (Porton 1995) have occurred in lemurs: 9 ruffed lemurs (range, 44 to 78 days; 1.84 percent failure rate), 8 ringtailed lemurs (range, 44 to 78 days; 4.0 percent failure rate), 1 brown lemur (*Eulemur fulvus*, at 43 days; 0.17 percent failure rate), and 2 bamboo lemurs (*Hapalemur griseus*, at 32 days and 41 days; 1.4 percent failure rate; another female conceived at day 128 postinjection but this is considered a zoo error). One red-bellied lemur (*Eulemur rubriventer*) conceived 12 days after her third Depo-Provera treatment, but it is believed the female did not receive the full dose (C. William, personal communication). Although limited data are available for nocturnal prosimians, 2 failures indicate an effective injection interval may be similar to that used in diurnal prosimians. A greater bush baby (*Otolemur crassicaudatus*) given a 2.5 mg/kg dose and a slow loris (*Nycticebus coucang*) given a 5 mg/kg dose conceived at 58 days postinjection. Taken together, these results suggest that 5 mg/kg every 40 days should be an effective dose for all prosimian species, with the exception of *Hapalemur* sp., for which a 30-day interval may be more appropriate.

Lee-Parritz and Roberts (1994) evaluated the efficacy of 3 and 10 mg/kg of Depo-Provera in one New World monkey species, the cotton-top tamarin. Four of four females treated with 3 mg/kg and three of four females treated with 10 mg/kg continued to exhibit ovarian cyclicity. The Contraception Database revealed two Depo-Provera failures in white-faced saki monkeys; one female conceived around day 44 postinjection at a 20 mg/kg dose and the other conceived on approximately day 54 at a 21 mg/kg dose. A cotton-top tamarin given a 96 mg/kg injection of Depo-Provera on day 10 postpartum conceived 78 days later. Finally, a pygmy marmoset (*Callithrix pygmaea*) conceived 64 days after a 10 mg/kg dose of Depo-Provera was administered. These data indicate an appropriate dose and treatment interval recommendation is still needed for New World monkeys, but a 20 mg/kg body weight dose at a 35- to 40-day interval may be effective in many species.

Mora and Johansson (1976) did some of the earliest work on Depo-Provera efficacy in Old World monkeys. In this study, 4 rhesus macaques (*Macaca mulatta*) were given 10 times the human female dose of Depo-Provera, but duration of efficacy was nevertheless the same as that for women. The authors found that despite the higher dose, the macaques had an ovulatory rise in plasma progesterone between 101 and 233 days following treatment, which is similar to the 90- to 245-day range found in women. Mora and Johansson (1976) suggest the results imply that macaques may metabolize the synthetic hormone more rapidly than humans, which may explain why they require a higher than expected dose. Support for this may be found in a study by Keiko et al. (1996), which reported that 15 Japanese macaque females treated with a single 7.5 mg/kg Depo-Provera injection were not prevented from conceiving, whereas 5 females treated with a single 19.4 mg/kg injection were effectively contracepted for the entire breeding season. The authors also reported that a single 15 mg/kg treatment of Depo-Provera effectively contracepted 5 long-tailed macaques (*Macaca fascicularis*) for 161 plus or minus 17 days as measured by plasma estradiol, progesterone, and luteinizing hormone (LH) levels and by menstrual cycles. In the Contraception Database, 2 Depo-Provera failures occurred in Japanese macaques. One female conceived 40 to 48 days after a 30 mg/kg dose and another conceived at approximately 13 days after a 42 mg/kg dose. That the 2 females conceived despite doses higher than those received by females in the study by Keiko et al. (1996) suggests the failures were likely the result of darting versus dose inadequacies. Finally, a single baboon conceived 3 months after an injection, but the dose was not reported. Additional research and experience are needed with Old World monkeys to determine an appropriate Depo-Provera dose and treatment interval.

Of the three failures of Depo-Provera in apes, all were in chimpanzees. Two failures occurred in the same chimpanzee female, one at 48 and the other at 60 days following Depo-Provera treatment. On both occasions the female was darted with the drug and may not have received the full dose. The third was a failure in a female that was contracepted with Depo-Provera while also receiving birth control pills. This particular female chimpanzee had previously experienced failures while also receiving different birth control pill formulations that have successfully contracepted other ape females, which suggests she is an extreme case best excluded from the analysis. Another female chimpanzee conceived 129 days after injection, exceeding the recommended injection interval by

39 days. Collectively, the results do indicate that the human contraceptive dose (150 mg every 90 days) should be effective in most female apes.

The Contraception Database contains records for 45 females that were taken off Depo-Provera and allowed to breed: 21 prosimians, 3 New World monkeys, 18 Old World monkeys, and 3 apes (see Table 8.2). The reversal rate for prosimians was 100 percent (number of injections ranged from 1 to 13). Of the 21 prosimian females, 74 percent conceived within one season while the remaining 26 percent of the females conceived by the second season. In 84 percent of the births at least one live-born infant was produced. All 3 New World monkeys, 1 spider monkey (*Ateles geoffroyi*) and 2 black howler monkeys (*Alouatta caraya*), conceived after Depo-Provera treatment and gave birth to live infants. However, the data in terms of number of individuals, dose, and duration of contraceptive use (1 or 2 injections) limit interpretation. Five of the 18 Old World monkeys (28 percent) became pregnant after Depo-Provera treatment. Within apes, a siamang (*Hylobates syndactylus*) conceived 15 months after a single injection, but the status of her infant was not reported. Two chimpanzee females conceived after 2 injections, 1 at 4 months and the other at 6 months following Depo-Provera treatment, and gave birth to healthy offspring.

The sample size for resumption of fertility data could be increased if we included the Depo-Provera failures that occurred because of inadequate information on effective injection intervals. The resulting births could be more appropriately categorized as resumption of fertility data [for example, the same milligram per kilogram (mg/kg) dose was effective at a shorter treatment interval]. If this criterion is accepted and if the summary is restricted to those species in which individuals are hand caught for treatment, thereby ensuring the entire dose was received, an additional 22 prosimians and 3 small-bodied New World monkeys have conceived following Depo-Provera treatment. This information is valuable not only in providing greater insight into appropriate treatment intervals but also in indicating that, in these females, treatment with Depo-Provera did not compromise their fertility. In no nonhuman primate species was time to reversal related to the total number of injections a female received, which agrees with results from the human literature (Schwallie and Assenzo 1974).

Twenty-three prosimians, 7 New World monkeys, 1 Old World monkey, and 3 apes (total, 34) were given one or more injections of Depo-Provera while already pregnant; 82 percent during the first, 15 percent during the second, and 3 percent during the third trimester (see Table 8.3). There were no reported problems during parturition. The majority (85 percent) of the females gave birth to live infants, including 24 of the 28 females treated during the first trimester, 4 of the 5 in the second, and the single female during the third trimester. Tarara (1984) surgically removed and necropsied six 100-day baboon fetuses from females treated with 150 mg Depo-Provera (7 to 8 mg/kg) during the first trimester. One fetus was dead but the five living fetuses were of normal weight, size, and developmental stage, indicating treatment did not have an adverse impact on fetal development.

Weight gain is likely with all progestins (AZA CAG 2004), as reported in baboons (Portugal and Asa 1995), chimpanzees (Bettinger et al. 1997), and common marmosets (Moehle et al. 1999) and as noted by some respondents to the Contraception Database Survey. Although no comparative data are available, subjective observations of Depo-Provera use in black lemurs indicate it may cause more weight gain than MGA implants (I. Porton, unpublished data). It should also be noted that many black lemur (a sexually dichromatic species) females treated with Depo-Provera experience a blackening of their fur (the male-typical color), especially on their face, head, neck, and areas that have been shaved (Porton 1995; Contraception Database).

PROGESTINS: ORAL Six oral progestin-only pills have been used in six species of nonhuman primates, including four ape species, one Old World monkey, and one prosimian (see Table 8.1). The four types of human birth control pills, often referred to as the minipill, were only used in apes (gibbon, chimpanzee, and gorilla). The products were Provera (medroxyprogesterone acetate; Upjohn), Ortho-Micronor (0.35 mg norethindrone), Microlut (0.03 levonorgestrel; Schering Ltd.), and Femulen (0.5 mg ethynodiol diacetate; Pharmacia). Minipill use in the apes ranged from 1 to 58 months. Two chimpanzees conceived after 16 and 19 months on the minipill Microlut, and a white-cheeked gibbon (*Hylobates leucogenys*) conceived after 5 months on Provera. The females were treated through all or part of their pregnancy (see Table 8.3); there were no reported complications during parturition and the infants were born alive.

The other two progestin-only pills (Ovaban: Schering-Plough, and Megace) contain megestrol acetate and are used solely in veterinary medicine. These products were used in two siamangs, four mandrills (*Mandrillus sphinx*), and one ruffed lemur. The mandrills were administered 60 mg Ovaban twice a week; the regimen for the siamang was not provided. Another siamang was treated with Megace at an initial dose of 20 mg per day for 3 weeks, after which the dose was reduced to 20 mg per week. The dose for the ruffed lemur treated with Magace was not provided. There were no reported failures with Ovaban (three mandrills treated for more than 4 years) or Megace (siamang treated for 83 months; ruffed lemur for four seasons). One mandrill was treated throughout pregnancy (see Table 8.3) and gave birth to a live infant.

ESTROGEN–PROGESTIN COMBINATIONS: IMPLANTS A trial to test the efficacy of a combination MGA–ethinyl estradiol implant was initiated through the AZA CAG to address the problem of implant size and loss in New World monkeys. The higher dose of MGA required to suppress estrus in New World monkeys requires a larger-diameter implant. An estrogen–progestin combination reduces the required MGA dose and thus the implant size. Three species of New World monkeys and one ape (an orangutan, a species that also experienced a high MGA implant loss rate) were treated with the combination implant. One squirrel monkey (*Saimiri sciureus*) and one white-faced saki monkey were implanted for 3 months each. Fourteen black-handed spider monkeys (*Ateles geoffroyi*) received the combination implant. Eight implants were removed within a year, two were in place for 2 years, and two implants were lost. Three orangutans were treated; one for 43 months, another for 36 months, and the third for 5 months. Four of the spider monkey females conceived between 2 and 14 months after implant insertion. As part of the research trial, endometrial biopsies were taken from 7 of the spider monkey females. After overstimulation of the uterine epithelium was discovered, even at noncontraceptive doses, all trials were ended (C. Asa and I. Porton, unpublished data).

ESTROGEN–PROGESTIN COMBINATIONS: ORAL The database contains records for seven nonhuman primate species that have been treated with birth control pills: one New World and one Old World monkey and five ape species (see Table 8.1). Twenty-one different birth control pills have been used. All combination birth control pills are composed of a progestin (typically norethindrone acetate or the more potent levonorgestrel) and an estrogen (typically ethinyl estradiol or mestranol) and differ by formulation (hormone amount or ratio) or only by brand; for example, Norinyl 1 + 50 (Syntex) and Ortho-Novum 1/50 (Ortho) contain the same formulation.

A total of 143 females were treated with combination birth control pills, including 3 spider monkeys, 3 mandrills, 3 gibbons, 80 chimpanzees, 30 gorillas, and 27 orangutans. The 3 spider monkeys are not included in this review because no details of their treatment are available. Most of the females, including the 3 mandrills and all the great apes, were treated at the human 28-day regimen of 21 days treatment and 7 days off (or placebo), with the exception of 2 chimpanzees that were treated continuously. Two siamangs were treated with a quarter pill of Ortho-Novum 1/50 and the third siamang was treated with a half pill. Fifty-six percent of the 78 completed contraceptive bouts were of less than 13 months use, 32 percent represented 13 to 48 months of use, and 12 percent represented more than 4 years of use. There are 86 ongoing bouts of oral birth control use, and if the species gestation period is subtracted from the ongoing bout, additional data on the efficacy of this method can be gained. Fifteen percent of the 86 ongoing bouts are less than 13 months, 53 percent are between 13 and 48 months, and 34 percent are bouts of greater than 4 years, including females that have been successfully contracepted for more than 8 years.

Ten failures have occurred in primates treated with combination birth control pills, including 5 orangutans, 3 chimpanzees, 1 gorilla, and 1 siamang. A siamang on the quarter-pill dose conceived at 6 months; the second siamang treated at this dose was lost to follow-up at 19 months. Among the great apes, 3 of the failures occurred in a single female chimpanzee that was treated with two different pill formulations (Modicon and Ortho Novum 1/50) and had an additional (third) failure when birth control pills were further supplemented with Depo-Provera (discussed previously). Excluding the chimpanzee just described, the siamang (low dose), and an orangutan female that missed five consecutive pills (Porton 1995), the birth control pill failure rate is 3.0 percent, which is higher than that reported for humans. However, the challenge with assessing "failures" in nonhuman primates is whether the resulting pregnancy was actually due to an inadequate dose or a failure of consumption. In several apes (in addition to the orangutan already described), it was suspected that a female did not consume her pill for 1 or more days. No one formulation has been shown to consistently fail.

Eleven apes were administered oral birth control while pregnant (see Table 8.3). In 10 cases, treatment was initiated during the first trimester and, although birth control was stopped as soon as it was realized the female was pregnant, 73 percent of the females were treated for more than half or through the remainder of their pregnancy. Ten infants were live born and 1 infant was stillborn. No problems with parturition or lactation were reported in CAG surveys. Rhesus monkey infants born to mothers experimentally treated with the birth control pill Norlestrin, at 20 to 100 times the human dose, showed no behavioral deficiencies when tested at ages 3 and 5 months (Golub et al. 1983).

Nineteen apes have been taken off oral birth control to allow reproduction and 14 have conceived. Time to reversal ranged from 1 to 44 months, but in 64 percent of the cases reversal time was 6 months or less. Time to reversal was not correlated to duration of birth control pill use. Goodrowe et al. (1992) analyzed urinary hormones from 2 western lowland gorillas and showed that within 7 days both females were effectively contracepted with Demulen 50 [1 mg ethynodiol diacetate, 50 micrograms (μg) ethinyl estradiol] for the duration of the 3-month trial, and both resumed cycling following treatment termination.

The effect of estrogen–progestin contraceptive treatment on the sexual swelling of chimpanzees has been investigated by Bettinger (1994) and Nadler et al. (1992). They found that whether and to what degree female chimpanzees exhibited a sexual swelling when treated with birth control pills was correlated with the dose, the estrogen to progestin ratio of the pill, and whether the pills were administered continuously throughout the month or stopped for the placebo phase. No placebo break in pill consumption resulted in the complete absence of swelling (Bettinger 1994; Contraception Database), and increased estrogen caused larger swellings (Nadler et al. 1992). Continued swelling, but not to full tumescence, occurred on different pill formulations, and interindividual variation in swellings has occurred in females treated with the same formulation (Bettinger and DeMatteo 2001). Birth control pills decreased but did not eliminate breeding behavior in chimpanzees with strong affiliative bonds but did end mating in weakly bonded pairs (Nadler et al. 1993). No systematic research has been done to evaluate the expression of sexual behavior and birth control pill use in the other ape species.

Protein Hormones

Because of limited availability of the GnRH agonist implants such as histrelin (The Population Council) and deslorelin (Suprelorin: Peptech Animal Health; see Chapter 3, Types of Contraception, for method details), and the prohibitively high cost of Lupron injections, there has been limited experience with GnRH agonists as a contraceptive alternative in zoo-housed primates. Initial trails with deslorelin implants involving five Goeldi's monkeys are ongoing and show promise. Deslorelin was tested in two ruffed lemur females; implants were inserted in November, and both females came into estrus within a week, after which ovulation was effectively suppressed for the entire breeding season. Information on the use of Lupron is limited to a female lion-tailed macaque that was successfully treated at monthly intervals for 6 months and a female gorilla that was successfully treated with two Lupron injections 4 months apart.

Immunocontraception

Zona pellucida (ZP) vaccines differ in the ZP immunogen that is used to stimulate the immune response. The challenge with ZP immunocontraception is to identify a vaccine that is effective but does not cause permanent sterility. Dunbar et al. (2002) summarized the results of seven studies of ZP immunocontraception in nonhuman primates. Ovarian dysfunction or pathology occurred with several of the immunogens that were tested whereas others proved effective and reversible. For example, Sacco et al. (1987) found that squirrel monkeys immunized with purified ZP3 using Freund's adjuvant resumed ovarian hormonal production at 10 to 15 months after treatment.

The Contraception Database contains records for nine species and 20 individual nonhuman primates that have been contracepted with porcine zona pellucida (PZP), including 2 prosimians (black lemur and ruffed lemur), 2 New World monkeys (owl monkey, *Aotus trivirgatus*, and cotton-top tamarin), 3 Old World monkeys (mandrill, patas monkey, *Erythrocebus patas*, and colobus), and 2 apes (chimpanzee and orangutan). Sample size for evaluation is even smaller because a patas monkey was already pregnant and a chimpanzee and 2 orangutan females were treated with a second contraceptive method while on PZP. Additionally, 2 females (a black lemur and a patas monkey) were not separated from reproductive partners during the initial PZP series and conceived around the time of the third injection. Interestingly, the black lemur female conceived again less than a month after giving birth in May, which is outside the normal breeding season for this species. Thirteen individuals were treated for 2 years and 2 individuals for 3 years. There were three failures. Two of the females gave birth to live offspring while the third miscarried. Four females (2 black lemurs, a ruffed lemur, and an owl monkey) developed deep tissue infections from the Freund's adjuvant; 2 of the females died. Because of the small sample size and low number of repetitions and reversal attempts, it is not yet possible to evaluate efficacy and reversibility of the vaccine PZP in primates.

Mechanical Methods

One barrier method has been used in nonhuman primate females, the intrauterine device (IUD). It has been tested as a contraceptive alternative in three apes species: chimpanzee, orangutan, and gorilla. In a 30-month study by Porteous et al. (1994), 21 chimpanzees, housed in a captive colony in Gabon, were contracepted with human IUDs. The IUDs were rejected in 3 females (14 percent), and another 4 (19 percent) conceived with the IUD in place. In another study by Gould and Johnson-Ward (2000) with 39 IUD-contracepted chimpanzees, a single female conceived with the IUD in place. Eight, mostly nulliparous females, lost their IUDs. Florence et al. (1977) successfully contracepted 1 zoo-housed orangutan with an IUD for 13 months. The female conceived 4 months after the IUD was removed. The authors noted removal was difficult, because the string had been shortened to prevent removal by the female. A single female gorilla in a European zoo was successfully contracepted with an IUD (Gerlofsma et al. 1994).

METHODS FOR MALES

Steroid Hormones

PROGESTINS Steroid hormones have not been used to contracept male nonhuman primates housed in zoos. Very few male primates (one lion-tailed macaque, three patas monkeys, one orangutan) have been treated with an MGA implant ($n = 2$) or Depo-Provera ($n = 3$) for behavioral but not contraceptive reasons. The limited sample size precludes evaluation (but see Chapter 12, Contraceptive Agents in Aggression Control).

Protein Hormones

Trials with histrelin or deslorelin as a possible management tool for aggression control in male lion-tailed macaques, black lemurs, and ringtailed lemurs have met with good success in the macaques (Norton et al. 2000; I. Porton, unpublished data; see Chapter 12, Contraceptive Agents in Aggression Control) and limited success in the lemurs (I. Porton, unpublished data; C. Williams, unpublished data). In terms of deslorelin effectiveness as a contraceptive, semen evaluation showed some but not all the male macaques were aspermic. Lemur fertility was not evaluated because of the health risks associated with electroejaculation in this taxon. Interestingly, most of the black lemur males treated with deslorelin exhibited some browning of their pelage, and a few males completely changed to the female color pattern of brown with white ear tufts. All pelage changes reversed following implant expiration (I. Porton, unpublished data). Four monthly treatments with Lupron Depot (TAP Pharmaceuticals) were given to a male spider monkey, but no fertility or behavioral results were provided.

Mechanical Methods

One barrier method, a silicone plug inserted into each vas deferens, has been used in nonhuman primate males. Vas plugs were placed in rhesus macaques and shown to be effective at blocking sperm passage for the 7- to 11-month trial period. All males ejaculated normal sperm within 2 months of vas plug removal (Zaneveld et al. 1988). The promising results led to testing the method on 11 species of zoo-housed primates (see Table 8.1). However, research was discontinued because of variation in vas deferens morphology and distensibility among the different species, which resulted in incomplete blockage (three chimpanzees, one baboon) or damage when inserting or removing the plugs (C. Asa and L. Zaneveld, unpublished data).

SUMMARY

The Contraception Database contains records for more than 2,000 contracepted primates and, combined with published research results, this information has begun to provide a better basis from which to develop contraception recommendations for nonhuman primates. The data indicate there are a number of contraceptives that are effective in the variety of nonhuman primates that are housed in captive facilities. Although the sample size for many species is negligible, some generalizations within the four broad taxonomic divisions can be made.

The two methods most widely used in prosimians, the MGA implant and Depo-Provera, are effective. Implant loss is a problem, particularly in the brown lemur and ringtailed lemur. Consequently, it is important for managers to confirm the presence of the implant whenever females are handled or when the presence of the implant is questioned. Depo-Provera is typically reliable in prosimians, because hand-catching these species provides greater assurance the appropriate dose is delivered. This method does, however, require the staff to be diligent in administering injections at regular intervals and necessitates accurate knowledge of the breeding season. Lemuridae appear to develop more extensive endometrial hyperplasia when administered MGA (see Chapter 5, Adverse Effects of Contraceptives); however, no direct link between infertility and progestin contraception has been found. Porton (in preparation) found 22 ruffed lemurs successfully conceived after one to six seasons of progestin contraception.

New World monkeys present the greatest contraceptive challenges. MGA has proved effective in the callitrichids. In two species with a relatively larger number of reversal attempts (golden lion tamarin and cotton-top tamarin), reversibility has been clearly shown (Wood et al. 2001; Contraception Database). The frequency of contraceptive reversals in the related golden-headed lion tamarin has been characterized as low by De Vleeschouwer et al. (2000a), but more data are needed (DeMatteo et al. 2002). Progestins are not advised for one species, Goeldi's monkey, as resulting uterine pathology has sterilized females treated with MGA (Murnane et al. 1996). Progestin contraception during pregnancy, a common practice in callitrichids because of their postpartum estrus, has not been shown to be a problem. Eighty-eight percent of the callitrichid females treated with MGA during pregnancy gave birth to at least one live infant. Less is known about effective contraception within the other New World monkey group, Cebidae. The failure rates among females contracepted with progestins have been relatively high, but the sample size has been small. Although some Cebidae females have been effectively treated with higher-dose MGA implants, the sample size is too small to allow conclusions. Initial positive results with a GnRH agonist, deslorelin, as an alternative to steroidal hormones for contraception in New World monkeys, are encouraging.

This Contraception Database review suggests MGA implants are highly effective in Old World monkeys, but the problem of implant loss is significant. Implant loss is particularly problematic in macaques, a genus renowned for their dexterity and exploratory nature. The recommended 5- to 7-day separation from group members following implant placement is especially important for these species, but separation can cause social problems. The use of implants therefore requires that animal managers develop reintegration strategies to minimize social discord. Depo-Provera, which does not require separation, could be a valuable alternative, especially if animals are trained to accept injections. Progestin-only and combination birth control pills have been effective in mandrills and may be useful with individuals that can be reliably hand fed. Data on reversibility are surprisingly limited, indicating that population growth rates are being carefully controlled in this primate group. In only 7 percent of the 579 MGA implant-treated females have implants been removed for reproduction, and more than 50 percent of these females have conceived. The data are still preliminary, because the results include females that have been in a breeding situation for less than a year.

Progestin implants and Depo-Provera are effective in apes, but implant loss is a problem, especially in orangutans and chimpanzees. Combination birth control pills are effective in great apes and are clearly preferred by many managers. This choice is likely because more intense individualized attention and interaction typifies ape husbandry, which makes the procedure of administering a daily pill realistic. Use of birth control pills also avoids the need to anesthetize females for implant insertion. Earlier database results (Porton et al. 1990; Porton 1997) questioned the advisability of contracepting orangutans with birth control pills because of the relatively high pregnancy rate among females on the pill. It was not known if failures were caused by dose or consumption issues. The current database contains additional information, including eight females that have been successfully treated with birth control pills for 25 to more than 80 months. This finding suggests the previous failures may not have been dose related and that human birth control pills are an appropriate contraceptive alternative for orangutans. The final decision, however, should be based on the caretaking staff's familiarity with the individual female's behavior and the likelihood that the staff will be able to confirm pill consumption.

The effect of contraception on primate behavior has been the focus of only a few behavioral studies (Steklis et al. 1983; Lee-Parritz and Roberts 1994; Portugal and Asa 1995; De Vleeschouwer et al. 2000b). No single pattern has emerged from these studies, and the results indicate understanding the effect of contraception on the sociosexual behavior of primates will require extensive research on a range of primate species in a variety of social and environmental situations. Female and male primates do not respond exclusively to hormonal cues. Rather, the plastic nature of primate behavior results in varied sociosexual interactions depending on the animals' physical environment, interindividual relationships, social milieu, and the situational context, in addition to the physiological state of the individuals (Fedigan 1992).

ACKNOWLEDGMENTS

The authors express their deep appreciation to the countless number of people who have devoted so much time and effort to responding to the CAG Contraception Surveys. We also thank Tammie Bettinger and Jeff French for helpful reviews of the manuscript.

REFERENCES

Abbott, D. H., J. Barrett, and L. M. George. 1993. Comparative aspects of the social suppression of reproduction in female marmosets and tamarins. In *Marmosets and Tamarins: Systematics, Behavior, Ecology*, ed. A. B. Rylands, 152–163. Oxford: Oxford University Press.

Asa, C. S., I. Porton, A. M. Baker, and E. D. Plotka. 1996. Contraception as a management tool for controlling surplus animals. In *Wild Mammals in Captivity*, ed. D. G. Kleiman, M. E. Allen, K. V. Thompson, and S. Lumpkin, 451–467. Chicago: University of Chicago Press.

American Zoo and Aquarium Association Contraception Advisory Group (AZA CAG) Recommendations. 2004. www.stlzoo.org/contraception.

Bardin, C. W., and I. Sivin. 1992. Norplant: the first implantable contraceptive. In *Contraception: Newer Pharmacological Agents, Devices, and Delivery Systems*, ed. R. Sitruk-Ware and C. W. Bardin, 23–29. New York: Dekker.

Bettinger, T. 1994. Effect of contraceptives on female chimpanzee genital swelling. In *AZA Regional Conference Proceedings*, 9–15, March 6–8, 1994, Oklahoma City, OK.

Bettinger, T. L., and K. E. DeMatteo. 2001. Reproductive management of captive chimpanzees: contraceptive decisions. In *The Care and Management of Captive Chimpanzees*, ed. L. Brent, 119–145. San Antonio: American Society of Primatologists.

Bettinger, T., D. Cougar, D. R. Lee, B. L. Lasley, and J. Wallis. 1997. Ovarian hormone concentrations and genital swelling patterns in female chimpanzees with Norplant implants. *Zoo Biol* 16 (3): 209–223.

Chaoui, N. J., and S. Hasler-Gallusser. 1999. Incomplete sexual suppression in *Leontopithecus chrysomelas*: a behavioural and hormonal study in a semi-natural environment. *Folia Primatol* 70:47–54.

Chrousos, G. P., D. Brandon, D. M. Renquist, M. Tomita, E. Johnson, D. L. Loriaux, and M. B. Lipsett. 1984. Uterine estrogen and progesterone receptors in an estrogen- and progesterone "resistant" primate. *J Clin Endocrinol Metab* 58:516–520.

Coe, C. L., A. Savage, and L. J. Bromley. 1992. Phylogenetic influences on hormone levels across the primate order. *Am J Primatol* 28:81–100.

DeMatteo, K. E., I. J. Porton, and C. S. Asa. 2002. Comments from the AZA contraception advisory group on evaluating the suitability of contraceptive methods in golden-headed lion tamarins (*Leontopithecus chrysomelas*). *Anim Welf* 11:343–348.

De Vleeschouwer, K., K. Leus, and L. Van Elsacker. 2000a. An evaluation of the suitability of contraceptive methods in golden-headed lion tamarins (*Leontopithecus chrysomelas*), with emphasis on melengestrol acetate (MGA) implants. (I) Effectiveness, reversibility and medical side-effects. *Anim Welf* 9:251–271.

De Vleeschouwer, K., K. Leus, and L. Van Elsacker. 2000b. An evaluation of the suitability of contraceptive methods in golden-headed lion tamarins (*Leontopithecus chrysomelas*), with emphasis on melengestrol acetate (MGA) implants. (II) Endocrinological and behavioural effects. *Anim Welf* 9:385–401.

Dunbar, B. S., G. Kaul, M. Prasad, and S. M. Skinner. 2002. Molecular approaches for the evaluation of immune responses to zona pellucida (ZP) and development of second generation ZP vaccines. *Reproduction Suppl* 60:9–18.

Fedigan, L. M. 1982. *Primate Paradigms*. Chicago: University of Chicago Press.

Florence, B. D., P. J. Taylor, and T. M. Busheikin. 1977. Contraception for a female Borneo orangutan. *J Am Vet Med Assoc* 171:974–975.

Gerlofsma, M. H., P. Zwart, and P. S. J. Klaver. 1994. Review of contraceptive methods in zoo mammals. *Verh Berl Erkrg Zootiere* 36:25–36.

Goldfoot, D. A., S. J. Wiegand, and G. Scheffler. 1978. Continued copulation in ovariectomized adrenal-suppressed stumptail macaques (*Macaca arctoides*). *Horm Behav* 11:89–99.

Golub, M. S., L. Hayes, S. Prahalada, and A. G. Hendrickx. 1983. Behavioral tests in monkey infants exposed embryonically to an oral contraceptive. *Neurobehav Toxicol Teratol* 5 (3): 301–304.

Goodrowe, K. L., D. E. Wildt, and S. L. Monfort. 1992. Effective suppression of ovarian cyclicity in the lowland gorilla with an oral contraceptive. *Zoo Biol* 11:261–269.

Gould, K. G., and J. Johnson-Ward. 2000. Use of intrauterine devices (IUDs) for contraception in the common chimpanzee (*Pan troglodytes*). *J Med Primatol* 29 (2): 63–69.

Gresl, T. A., S. T. Baum, and J. W. Kemnitz. 2000. Glucose regulation in captive *Pongo pygmaeus abeli*, *P. p. pygmaeus* and *P. p. abeli* × *P. p. pygmaeus* orangutans. *Zoo Biol* 19 (3): 193–208.

Hayssen, V., A. van Tienhoven, and A. van Tienhoven. 1993. *Asdell's Patterns of Mammalian Reproduction: A Compendium of Species-Specific Data*. Ithaca: Comstock.

Keiko, S., T. Yuji, M. Fusako, and N. Masumi. 1996. Suppression of ovarian function and successful contraception in macaque monkeys following a single injection of medroxyprogesterone acetate. *J Reprod Dev* 42 (2): 147–155.

Lee-Parritz, D. E., and E. Roberts. 1994. Efficacy of hormonal contraception in cotton-top tamarins (*Saguinus oedipus*). In *AZA Regional Conference Proceedings*, 387–390, March 6–8, 1994, Oklahoma City, OK.

Michael, R. P., and D. Zumpe. 1971. Patterns of reproductive behavior. In *Comparative Reproduction of Nonhuman Primates*, ed. E. S. E. Hafez, 205–242. Springfield: Thomas.

Moehle, U., M. Heistermann, A. Einspanier, and J. K. Hodges. 1999. Efficacy and effects of short-and medium-term contraception in the common marmoset (*Callithrix jacchus*) using melengestrol acetate implants. *J Med Primatol* 28 (1): 36–47.

Mora, G., and E. D. B. Johansson. 1976. Plasma levels of medroxyprogesterone acetate (MPA), estradiol and progesterone in the rhesus monkey after intramuscular administration of Depo-Provera. *Contraception* 14 (3): 343–350.

Murnane, R. D., J. M. Zdziarski, T. F. Walsh, M. J. Kinsel, T. P. Meehan, P. Kovarik, M. Briggs, S. Raverty, and L. G. Phillips. 1996. Melengestrol acetate-induced exuberant endometrial decidualization in Goeldi's marmosets (*Callimico goeldii*) and squirrel monkeys (*Saimiri sciureus*). *J Zoo Wildl Med* 27 (3): 315–324.

Nadler, R. D., J. F. Dahl, D. C. Collins, and K. G. Gould. 1992. Hormone levels and anogential swelling of female chimpanzees as a function of estrogen dosage in a combined oral contraceptive. *Proc Soc Exp Biol Med* 201:73–79.

Nadler, R. D., J. F. Dahl, K. G. Gould, and D. C. Collins. 1993. Effect of an oral contraceptive on sexual behavior of chimpanzees (*Pan troglodytes*). *Arch Sex Behav* 22: 477–500.

Norton, T. M., L. M. Penfold, B. Lessnau, W. Jochle, S. I. Staaden, A. Jolliffe, J. E. Bauman, and J. Spratt. 2000. Long acting deslorelin implants to control aggression in male lion-tailed macaques (*Macaca silenus*). In *Proceedings, American Association of Zoo Veterinarians*, September 17–21, 2000, New Orleans, LA.

Porteous, I. S., N. I. Mundy, and C. Grall. 1994. Use of intrauterine devices as a means of contraception in a colony of chimpanzees (*Pan troglodytes*). *J Med Primatol* 23 (6): 355–361.

Porton, I. 1995. Results for primates from the AZA contraception database: species, methods, efficacy and reversibility. In *Proceedings, Joint Conference of the American Association of Zoo Veterinarians, Wildlife Disease Association, and American Association of Wildlife Veterinarians*, 381–395, August 12–17, 1995, East Lansing, MI.

Porton, I. 1997. Birth control options. In *Orangutan Species Survival Plan Husbandry Manual*, ed. C. Sadaro, 36–44. Chicago: Chicago Zoological Park.

Porton, I., C. Asa, and A. Baker. 1990. Survey results on the use of birth control methods in primates and carnivores in North American zoos. In *Annual Conference Proceedings of the American Association of Zoo Veterinarians*, 489–497, September 23–27, 1990, Indianapolis, IN.

Portugal, M. M., and C. S. Asa. 1995. Effects of chronic melengestrol acetate contraceptive treatment on perineal tumescence, body weight, and sociosexual behavior in hamadryas baboons (*Papio hamadryas*). *Zoo Biol* 14:251–259.

Richard, A. F. 1985. *Primates in Nature*. New York: Freeman.

Sacco, A. G., E. C. Yurewicz, M. G. Subramanian, E. C. Yurewicz, and W. R. Dukelow. 1987. Ovaries remain functional in squirrel monkeys (*Saimiri sciureus*) immunized with porcine zona pellucida 55 000 macromolecule. *Biol Reprod* 36:481–490.

Schwallie, P. C., and J. R. Assenzo. 1974. The effect of depo-medroxyprogesterone acetate on pituitary and ovarian function and the return of fertility following its discontinuation: a review. *Contraception* 10 (2): 181–197.

Steklis, H. D., G. S. Linn, S. M. Howard, A. Kling, and L. Tiger L. 1983. Progesterone and socio-sexual behavior in stumptailed macaques (*Macaca arctoides*): hormonal and socio-environmental interactions. In *Hormones, Drugs and Social Behavior in Primates*, ed. H. D. Steklis and A. Kling, 107–135. New York: Spectrum.

Tarara, R. 1984. The effect of medroxyprogesterone acetate (Depo-Provera) on prenatal development in the baboon (*Papio anubis*): a preliminary study. *Teratology* 30 (2): 181–185.

Van Elsacker, L., M. Heistermann, J. K. Hodges, A. DeLaet, and R. F. Verheyen. 1994. Preliminary results on the evaluation of contraceptive implants in golden-headed lion tamarins, *Leontopithecus chrysomelas*. *Neotrop Primates* 2 (suppl): 30–32.

Wood, C., J. D. Ballou, and C. S. Houle. 2001. Restoration of reproductive potential following expiration or removal of melengestrol acetate contraceptive implants in golden lion tamarins (*Leontopithecus rosalia*). *J Zoo Wildl Med* 32 (4): 417–425.

Zaneveld, L. J. D., J. W. Burns, S. Beyler, W. Depel, and S. Shapiro. 1988. Development of a potentially reversible vas deferens occlusion device and evaluation in primates. *J Fertil Steril* 49 (3): 527–533.

MARILYN L. PATTON, WOLFGANG JÖCHLE,
AND LINDA M. PENFOLD

9
CONTRACEPTION IN UNGULATES

Contraceptive use in ungulates began more recently than in some other taxa, especially felids and primates. However, issues common to the management of other species have become increasingly important in ungulates as well. In particular, space limitations, the need for sound genetic management, and the desire to maintain animals in natural social groups dictate that reproduction be controlled. Most ungulates are herd animals, with one or a few males consorting with a group of females, typically including mothers and their offspring (Leuthold 1977). In free-ranging situations, young animals can disperse or new males may enter a herd. However, in captive conditions, where the herd male or males are likely the sires of the young females, there are continuous opportunities for inbreeding. Conscientious managers rotate herd sires to avoid inbreeding or attempt to predict puberty and move young animals to other groups or institutions as they reach adulthood (imitating dispersal), but this strategy often fails when puberty cannot be anticipated or a new home found quickly enough. Yet, even when it succeeds and the young adult is placed with nonrelated individuals, it is likely that reproductive rate will eventually overwhelm resources and available space. Given that space limitations exist and that genetic variation is desirable, it is prudent to manage animals through contraception.

The American Zoo and Aquarium Association (AZA) committees that advise member institutions regarding allocation of space and animal breeding assignments (for example, Species Survival Plans, Taxon Advisory Groups, and Population Management Plans, explained in the introduction to Part III: The Application) have added increasingly more ungulate species to the lists of taxa to be managed using contraception. Other regional zoo associations such as those in

Europe and Australasia have similar programs. As is the case for other taxa represented in Part III, data on ungulate contraception come primarily from AZA zoos, but a growing number of institutions from around the world are now contributing Contraception Surveys.

Detailed explanations of the contraceptive methods are presented in Chapter 3 and are not repeated here. Unless otherwise indicated, information in this chapter has been derived from the AZA Contraception Database. The Database, as of December 2001, indicated that contraceptives had been used in 92 species of ungulates more than 1,000 times. Perhaps not surprisingly, most of those ($n = 973$) were administered to females, because more methods are available for and have been tested in females. As for other taxa, those methods ranged from steroid hormones (primarily progestins) through gonadotropin-releasing hormone (GnRH) agonists to zona pellucida vaccines.

METHODS FOR FEMALES

Steroid Hormones

PROGESTINS: MELENGESTROL ACETATE IMPLANTS As has been seen for carnivores and primates, the implant containing melengestrol acetate (MGA) (available from E. D. Plotka) is the most commonly used contraceptive in captive ungulates (Table 9.1). A likely reason for its popularity is that it is distributed free of charge to zoos in the United States. In addition, the estimated duration of efficacy of an MGA implant is at least 2 years, in contrast to a Depo-Provera injection (Pharmacia and Upjohn), which is effective for only 1 to 3 months. MGA implants are reported for eight ungulate families, with 70 percent used in Bovidae, 13 percent in Cervidae, 8 percent in Camelidae, 4 percent each in Hippopotamidae and Tapiridae, and less than 1 percent for Suidae, Tayassuidae, and Giraffidae. None were reported for the Equidae, because MGA is not effective in that family. Of the 254 completed MGA implant contraceptive bouts for ungulates, 32 percent were in place for less than 12 months and 36 percent for 13 to 24 months.

As with most other mammals, MGA implants have been extremely effective in ungulates. The rate of confirmed failure (that is, the implant was present at the time of conception) was 3.6 percent; however, if cases are included in which the implant may have been lost or if the dose was later determined to be too low, the failure rate was 5.6 percent. One species, hippos (*Hippopotamus amphibius*), accounted for most of the ungulate MGA implant failures, with 40 percent

Table 9.1
Contraceptive methods, number of treatments, and number of individuals for ungulates[a]

Method	Number of delivered treatments	Number of ongoing treatments	Number of species	*n*
Steroid hormones:				
Progestin implants:				
MGA	386	97	67	278♀/20♂
Norplant	1	0	1	1♀
Jadelle	5	4	2	5♀
Norgestomet	1	0	1	1♀
Progestin injections:				
Depo-Provera	534	0	24	87♀/6♂
Oral progestins:				
Ovaban	1	0	1	1♀
MGA	1	0	1	1♀
MGA feed	13	0	1	13♀
Protein hormones:				
Deslorelin implant	10	6	4	10♀
Lupron Depot	17	0	3	1♀/5♂
GnRH implant	6	0	1	6♂
Immunocontraception:				
PZP	520		39	138♀
Mechanical:				
Vas plugs	4	0	2	4♂

MGA, melengestrol acetate; GnRH, gonadotropin-releasing hormone; PZP, porcine zona pellucida.

[a]Data from the Contraception Database.

confirmed failures, but a rate as high as 45 percent if those with possible lost implants are included. If hippos are excluded from the calculation, the overall failure rate for ungulates is 2 percent.

Although the rate of implant loss at 7.4 percent was not exceptionally high, it did constitute the major problem reported. However, steps can be taken to both reduce the chances of loss and facilitate monitoring the presence of the implant. These procedures include using sterile technique during insertion to minimize the risk of infection and placing a microchip transponder inside the implant so its location can be confirmed (see Chapter 4, Assessing Efficacy and Reversibility).

Especially because contraception in ungulates is rather recent, there have been too few attempts at reversal for proper analysis. In addition, many reversal attempts are ongoing, that is, have not been extended for sufficient time to judge success. For those that have already culminated in conception, summaries of duration of MGA implant use and time elapsing until conception are presented in Table 9.2. Seven females from six species have conceived following implant removal. Prior duration of contraception ranged from 1.5 to 29 months, and latency to conception following implant removal ranged from 1 to 31 months, similar to the individual variability seen in other taxonomic groups. Pregnancies in three female common eland with implants in place occurred after 3 years, which is 1 year past the predicted duration of efficacy.

A study to evaluate the effect of MGA on parturition in captive white-tailed deer (*Odocoileus virginianus*) showed that MGA delayed or prevented normal parturition in all five treated animals (Plotka and Seal 1989). Because of those results, the Contraception Advisory Group (CAG) has recommended against using MGA implants or other progestins during pregnancy. However, it is not always possible to diagnose pregnancy at the time of implant insertion, which means that pregnant females are sometimes implanted. Inspection of the Contraception Database (Table 9.3) reveals that seven individuals from five species were implanted with MGA during gestation. Surprisingly, parturition was normal in four of the five females with implants still in place at the end of gestation. Three females did experience problems that may have been caused by the MGA implant: a Siberian ibex (*Capra sibirica*) required a cesarian section, a muntjac (*Muntiacus reevesi*) required intervention 3 weeks after her expected due date, and another muntjac died of septicemia, although the gestational age of the mummified fetus was not reported (see Table 9.3). The reason some females had apparently normal parturitions while others experienced sometimes fatal complications (cf. study by Plotka and Seal 1989) is unclear, but this variation may be related to dosage; that is, there may be a dosage range that is high enough to prevent conception but below the level that might block uterine contractions. Thus, unless a safe dosage can be determined for each species, the recommendation against treating pregnant ungulates with MGA or any other progestin is the more cautious approach.

In general, MGA has been shown to be both safe and effective in several ungulate species (Zimbelman et al. 1970; Munson 1993), and the most common problems associated with MGA implant use have been implant loss and migration from the implantation site. However, abscess formation sometimes occurs at the MGA implant insertion site. Less commonly, single reports of hydrometra (big-

Table 9.2
Successful reversals for ungulates treated with contraceptives[a]

Method	Genus	*n*	Mean (range) duration of contraceptive use, months	Mean (range) time to conception, months[b]
MGA implants	*Addax nasomaculatus*	2	25.5 (22–29)	7 (4–10)
	Tragelaphus strepsiceros	1	10	4
	Taurotragus oryx	1	28	0[c]
	Ovis canadensis	2	18 (17–19)	24.5 (18–31)
	Oryx leucoryx	1	1.5	24
	Lama guanicoe	2	18 (17–19)	8.5 (1–16)
	Cervus duvauceli	1	12	1
Depo-Provera	*Giraffa camelopardalis*	1	1.5	1
	Hemitragus jemlahicus	1	3[d]	<1
	Alces alces	2	3	11
	Hippopotamus amphibius[e]	1	27	3

MGA, melengestrol acetate.

[a]Data from the Contraception Database.

[b]Conception data calculated by deducting gestation length from birth date using gestation lengths from Hayssen et al. (1993).

[c]Female became pregnant after predicted duration of implant effectiveness had passed, although implant was still in place.

[d]Assumed duration of efficacy was 3 months.

[e]Three females conceived following 500-mg dose of Depo-Provera, but as this may not be a contraceptive dose, it is not included in the reversal data.

horn sheep, *Ovis canadensis*; urial sheep, *O. vignei*; greater kudu, *Tragelaphus strepsiceros*), pyometra (barasingha, *Cervus duvauceli*; sable antelope, *Hippotragus niger*; greater kudu; Baird's tapir, *Tapirus bairdii*; Brazilian tapir, *T. terrestris*; nilgiri tahr, *Hemitragus hylocrius*), and mucometra (common eland, *Taurotragus oryx*; nilgiri tahr) have been associated with MGA administration.

PROGESTINS: OTHER IMPLANTS Data are limited on the use of other types of progestin implants in ungulates. Three types were reported for Bovidae, including Jadelle implants in four markhor (*Capra falconeri*) that are ongoing bouts and in one mountain goat (*Oreamnus americanus*) that failed or was lost (that female was reported to have died of septic uterine metritis, although the potential influence of the contraceptive cannot be determined). In addition,

Table 9.3
Successful parturition in ungulates treated with contraceptives during pregnancy[a]

Method/species	*n*	Trimester contraception started	Trimester contraception stopped	Successful parturition (*n*)
MGA implant				
Antidorcas marsupialis	1	3	—	1
Bos taurus	1	1	—	1
Bos frontalis	2	1	—	2
Capra sibirica	1	1	—	0[b]
Muntiacus reevesi	1	?	—	0[c]
Muntiacus reevesi	1	1	—	0[d]
Depo-Provera				
Camelus bactrianus	1	3	3	1
Hippopotamus amphibius	1	1	3	1[e]
Giraffa camelopardalis	1	1	—	0[f]
Giraffa camelopardalis	1	2	—	1
Muntiacus reevesi	1	2	3	1
Muntiacus reevesi	1	2	3	1
Ovibos moschatus	1	3	—	1
Deslorelin				
Kobus megaceros	3	1	—	3

Continued on next page

Norplant was used in one mountain goat, but efficacy cannot be ascertained because the implant was lost. Norgestomet incorporated into a biobullet was reported to be effective in another mountain goat for 19.5 months.

Norgestomet implants also have been used in white-tailed deer (DeNicola et al. 1997) and black-tailed deer (*Odocoileus hemionus*; Jacobsen et al. 1995). The contraceptive effectiveness of the implants for white-tailed deer was 92 percent, 100 percent, and 100 percent for doses of 42, 28, and 21 mg, respectively. Lower doses (14 mg) were less effective (48 percent; DeNicola et al. 1997). A single biobullet containing norgestomet (42 mg) prevented conception in black-tailed deer for one breeding season (Jacobsen et al. 1995).

Silastic implants containing levonorgestrel (200 mg) were ineffective at suppressing estrus and ovulation in white-tailed deer, the only ungulate species re-

Table 9.3 continued

Method/species	*n*	Trimester contraception started	Trimester contraception stopped	Successful parturition (*n*)
PZP				
Addax nasomaculatus	3	1	—	3
Bison bison	1	1	—	1
Bison bison	3	2	—	3
Bubalus depressicornis	1	1	—	1
Capra falconeri	1	2	—	1
Capra ibex	1	2	—	1
Cervis duvauceli	1	3	—	1
Elaphurus davidianus	1	3	—	1
Equus burchelli	2	1	—	2
Gazella dama	1	1	—	1
Giraffa camelopardalis	5	1	—	5
Giraffa camelopardalis	1	3	—	1
Hippopotamus amphibius	1	1	—	1
Hemitragus hylocrius	1	2	—	1
Muntiacus reevesi	1	1	—	1
Tragelaphus strepsiceros	1	1	—	1
Tragelaphus strepsiceros	1	2	—	1
Tragelaphus strepsiceros	1	3	—	1

MGA, melengestrol acetate; PZP, porcine zona pellucida.

[a]Data from the Contraception Database.

[b]Female died during caesarean section.

[c]Female died of septicemia; necropsy showed mummified fetus; estimated age of fetus not provided.

[d]Female 3 weeks overdue; removed mummified fetus; could not find implant.

[e]Depo-Provera dose was inadequate to prevent conception; female received three treatments during pregnancy; no problems during parturition.

[f]Third-trimester abortion.

ported treated with this progestin, possibly because the dose was inadequate (Plotka and Seal 1989).

PROGESTINS: INJECTIONS Depo-Provera injections (medroxyprogesterone acetate, MPA: Pharmacia and Upjohn) represent the second most commonly used progestin contraceptive for ungulates, reported in 7 families, 24 species, and 87 individuals. Of those, 27 percent were giraffes (*Giraffa camelopardalis*), 17 percent hippos, and the rest, 56 percent, an assortment of other species. There were 529 reports of separate injections, because the duration of efficacy is rather short (1 to 3 months, depending on dosage and species; see Part III, The Application, for more information on calculating dosages). Most of those injections involved 2 species, giraffes (41 percent) and hippos (33 percent), with another 13 percent administered to bovids, 6 percent to camelids, 3 percent to suids, and 2 percent each to cervids and tapirs. None were given to equids, as MPA seems not to be effective in that taxon. Of the 93 treatment periods, 55 percent comprised 1 or 2 injections and 19 percent comprised 11 to 31 injections. Long-term treatment was primarily directed to giraffes and hippos.

Depo-Provera often is selected for larger species in which anesthesia, which is required for insertion of MGA implants, can be an unacceptable risk. For this reason, a substantial amount of information is available concerning MPA treatment in hippos and giraffes, two species that present particular challenges for anesthesia.

The only failures were recorded for hippos. These data included three failures at a dose of 500 mg at 4 to 6 weeks after injection and one failure using 800 mg at 11 to 12 weeks. These failures may be explained by the finding of Graham and coworkers (2002), who monitored fecal progestins of five hippos each treated once with 0.8 g MPA. Progestin levels returned to pretreatment values by day 100 (range, 70 to 116) after injection, and one of these females conceived 100 days after injection. These results suggest that when a proper dosage regimen is used, Depo-Provera can be a very effective form of contraception in ungulates, excepting equids.

Births following MPA treatment occurred in giraffe, moose (*Alces alces*), Himalayan tahr (*Hemitragus jemlahicus*), and three hippos (see Table 9.2). Six female ungulates received Depo-Provera injections during pregnancy, but treatment ceased for three females when they were determined to be pregnant (see Table 9.3). Of these, one musk ox and one giraffe gave birth normally but another giraffe aborted in the third trimester. The hippos conceived 4 months after receiving three 500-mg injections, 7 months after two 500-mg injections, and 5 months after twelve 500-mg injections, respectively. Because they conceived during treatment, it is clear that 500 mg was not an effective contraceptive dose. Thus, because the contraception was not adequate, the subsequent births are not truly evidence of reversals.

Limited information is available regarding the safety of MPA in ungulates. At necropsy, a giraffe treated with MPA for 26 months was found to have inactive ovaries and uterine pathology, including glandular atrophy, mild endometritis, and mucometria. These uterine changes also have been observed in other progestin-treated ungulates (Munson 1993) and were attributed to either MPA treatment or advanced age.

PROGESTINS: ORAL There are only two records of progestin pill use in ungulates. One was a hippo treated successfully for 19 months with Ovaban and another, treated for 88 months with what was referred to as an MGA pill, that conceived during treatment. This treatment was likely also megestrol acetate (Ovaban), not melengestrol acetate, which to our knowledge has not been produced in pill form.

Because MGA is orally active, it can be milled into feed, which is a good way to manage mixed-sex groups of ungulates. However, subordinate animals, whose access to food may be limited by more dominant conspecifics, may not receive sufficient MGA to prevent unwanted pregnancies (P. Calle, personal communication). In a study in Chinese goral (*Naemorhedus goral*; $n = 8$), females treated with MGA milled in feed (0.8 g/day) during one reproductive season (August 25 to April 5) did not conceive. Reversibility was demonstrated during the following year when six of the eight conceived, including one subadult female that attained maturity and bred successfully (Patton et al. 2000). The status of the remaining two animals is not known because they were lost to follow-up after transfer to another institution. Possible physiological effects on the five male goral in the group that also consumed MGA-treated feed were not measured (Patton et al. 2000). Likewise, pregnancy was prevented in a mixed herd of barasingha, axis deer (*Cervus axis*), and blackbuck (*Antilope cervicapra*) that received 0.7 mg MGA/lb body weight milled in feed per day for 18 months (Raphael et al. 1992).

Altrenogest (Regu-mate; Hoechst-Roussel), which also is an orally active progestin, was not effective as a contraceptive when administered to a herd of 10 female and 14 male Himalayan tahr over a 6-month period (Raphael et al. 1992).

Protein Hormones

Gonadotropin-releasing hormone (GnRH) agonists have been shown to suppress estrous cycles in domestic ungulates such as cattle (Herschler and Vickery 1981; D'Occhio et al. 1996, 2002; D'Occhio and Apsden 1999), sheep (McNeilly and Fraser 1987), and pigs (Brussow et al. 1996). In cattle, GnRH agonists have been shown to be safe and effective in achieving controlled, reversible suppression of estrous cycles (D'Occhio et al. 1996), and pregnancies were prevented for 200 and 300 days with 8- and 12-mg implants, respectively (D'Occhio et al. 2002). The GnRH agonist leuprolide suppressed fertility in captive female wapiti (*Cervus elaphus nelsoni*) for a breeding season in doses of 45, 90, and 180 mg administered as a subcutaneous bioimplant (Baker et al. 2002). Five female Zambesi lechwes (*Kobus lechwe lechwe*) received 9-mg implants containing deslorelin acetate, another GnRH agonist. Three of the five were later confirmed to have been approximately 2 months pregnant at the time of implant placement, but all three delivered healthy calves. Subsequently, one conceived 11 months after implant placement, another did not give birth again for at least 19 months, and the third died of unrelated reasons. The two other lechwes were not pregnant at implant placement. One of these gave birth every year for the 4 years before deslorelin treatment, conceived 22 months following implant placement, and subsequently had a healthy calf. These circumstances suggest contraception was successful for almost 2 years but then safely reversed. The second of these females did not calve for at least 22 months after deslorelin treatment (M. Patton, unpublished data).

In another study, 24 female white-tailed deer were divided into four groups and treated with deslorelin implants (12 mg) subcutaneously (SC) or intramuscularly via a biobullet delivered by remote injection. Nonpregnant does were treated before onset of the breeding season in October and pregnant does in May, outside the season. For deer implanted SC in October, the contraceptive was 100 percent effective during the subsequent breeding season and 25 percent effective in the following year. For those treated with biobullets in October, contraceptive efficacy was 100 percent during the first year but zero in the subsequent year. Treatment with SC implants in May prevented conceptions in 100 percent of the animals during the subsequent breeding season but not year 2 (0 percent). Biobullet treatment in May was completely ineffective during the subsequent breeding season (Tart 1999).

Reports to the Contraception Database indicate that deslorelin contraception has been successful in 10 individuals from four ungulate species, including Nile lechwe (*Kobus megaceros*), European bison (*Bison bonasus*), sitatunga (*Tragelaphus spekei*), and bongo (*T. eurycerus*). Six of these bouts are ongoing, and 3 of the females were pregnant when implanted. A female hippo was injected once with Lupron, then switched to another form of contraception, so efficacy cannot be evaluated. Three Nile lechwes were treated with deslorelin during early pregnancy, but all gave birth normally.

Immunocontraception

ZONA PELLUCIDA VACCINES Porcine zona pellucida (PZP) vaccine has been reported for 9 ungulate families, 39 species, and 138 individual ungulates. The families include Camelidae, Tapiridae, Suidae, Cervidae, Bovidae, Giraffidae, Equidae, Hippopotamidae, and Proboscidae. The largest sample sizes are found in giraffes, 30 females, followed by Grevy's zebra (*Equus grevyi*), 15 females. Only 1 individual was treated in 18 other species. PZP vaccination was the only method used in captive Asian elephants (1 female) and equids (25 females).

Animals received either (1) an initial inoculation of PZP with modified Freund's adjuvant (MFA) followed by a booster of PZP with Freund's incomplete adjuvant (FIA), (2) three inoculations of PZP with FIA, spaced about 2 weeks apart, or (3) an initial inoculation of PZP with Freund's complete adjuvant (FCA) followed by a single booster of PZP with FIA 2 to 3 weeks later. In all cases a single annual booster inoculation of PZP with FIA was given at 7- to 12-month intervals, depending on the species. When technical errors were eliminated (incomplete injections, initiation of treatment after conception, inadequate needle length, failure to give booster inoculations on schedule, and access to a male before initial inoculation series was complete), efficacy ranged from 80 to 100 percent (Kirkpatrick et al. 1996; Frank and Kirkpatrick 2002). A problem with assessing results is that not all treatments included the complete vaccination series; efficacy is not assured until at least one or in some cases two booster injections are given. In several cases, Depo-Provera injections also were given with the PZP (especially during the first or second injection in the series), which confounds the results. In addition, nine females were pregnant when PZP treatment was initiated.

Nine of the reported failures were caused by introduction of the male before completion of the booster series. These lapses represent failures of administration or management, not of the contraceptive method. Ten contraceptive method failures were reported, although delivery of the entire dose could not be confirmed in some cases. Also, the duration of efficacy may be shorter in some species, requiring booster injections at more frequent intervals.

Data from PZP-treated ungulates over the past decade have demonstrated that not all mammals mount similar immune responses, as reflected by differences in anti-PZP antibody titers following identical treatment regimens. For example, after a series of three PZP inoculations (with FIA, over a 6-week period), Himalayan tahr produced antibody titers that ranged from 94 to 100 percent of reference serum standards, whereas muntjac produced antibody titers of 19 to 28 percent following the same treatment protocol (Kirkpatrick et al. 1996). Three bighorn sheep receiving the two-inoculation protocol of PZP plus FCA followed by PZP plus FIA remained infertile for 2 years without subsequent annual booster inoculations (Frank and Kirkpatrick 2002). One of the three females gave birth 3 and 4 years after PZP treatment, a second female gave birth 4 years after treatment, and the third had not returned to fertility 5 years after treatment. Among eight mountain goats given the two-inoculation protocol of PZP plus FCA followed by PZP plus FIA, one gave birth 3 years later and another 4 years later, but the rest had not produced young 4 years after inoculation. Thus, in goats and sheep an initial two-inoculation PZP series provided effective contraception for 2 years or longer, suggesting that reversibility might be a problem. In contrast, eight of nine ibex that received the three-inoculation treatment with FIA plus annual boosters of PZP plus FIA were infertile for 3 years, but one became pregnant during the third year of treatment (Frank and Kirkpatrick 2002).

Deleterious side effects associated with PZP contraception appear to be rare and minor in nature. For example, abscesses have historically been a problem associated with the use of FCA, but the published reports are almost exclusively focused on laboratory animals (Smith et al. 1992). When FCA is administered exclusively in the hip or gluteal muscles, abscesses occur at a rate of less than 1 percent (Turner and Kirkpatrick 2002; Deigert et al. 2003); they tend to be no larger than about 25 mm and drain without untoward effects. Another concern with the use of FCA is the possibility of false-positive tuberculosis (TB) tests following exposure to the adjuvant. The adjuvant contains killed cell-wall fragments of *Mycobacterium tuberculosis*, which provoke the positive TB tests but also are what make the adjuvant so effective. Thus, its use must be restricted to animals that are not to be moved between facilities (which requires TB testing to ensure absence of disease), or to equids, for which there is no reliable TB test.

The problems associated with FCA prompted the development of the three-inoculation protocol using FIA, which does not contain *M. tuberculosis* and thus does not elicit a false-positive TB test result. The initial inoculation with FIA elicits a weaker antibody response than FCA, but as the subsequent two boosters are given, antibody titers rise to effective contraceptive levels (Deigert et al. 2003). To alleviate this problem, the use of MFA was initiated with PZP inoculations. This adjuvant has the same oil base as FCA but utilizes killed cell-wall fragments of *Mycobacterium butyricum*, which appears to elicit strong antibody responses but cannot cause false-positive TB test results (Deigert et al. 2003). When these adjuvants are used, a small granuloma forms at the injection site in most cases; this is not an abscess but a small mass of scar tissue just under the skin, analogous to the scar left behind on the arm of individuals inoculated for smallpox. These granulomas are about the size of small marbles and do not cause any untoward health effects even when monitored for extended periods of time (J. Kirkpatrick, personal communication).

Among captive wild animals, treatment during pregnancy has not interfered with the progress of gestation or the health of offspring. Of 27 individuals from 18 species receiving PZP inoculations during the first, second, or third trimester, as reported to the Contraception Database, all pregnancies culminated in a successful parturition (see Table 9.3). The outcome also was normal in free-ranging African elephants (*Loxodonta africana*) treated with PZP during pregnancy (Delsink et al. 2002).

However, even though most ungulates have demonstrated no serious health problems as a result of PZP vaccine use, histopathological evaluations of PZP-treated animals are needed to eliminate concerns about the potential of permanent ovarian damage. In some species immunization with PZP can elicit an immune response that interferes with normal follicular development (Dunbar et al. 2001). Thus far, histological evaluations of PZP-treated white-tailed deer and feral domestic horses reveal no ovarian damage after 3 years of continuous application (McShea et al. 1997). Finally, because ovarian activity is not suppressed in PZP-treated animals, they may be susceptible to the pathological effects of continuous steroid exposure (Munson 2002).

Immunocontraception

ANTIGONADOTROPIN-RELEASING HORMONE VACCINES Immunization against GnRH has been shown to prevent pregnancy by blocking follicle growth and ovulation and has been successful in a range of domestic ungulates, such as sheep (Clarke et al. 1979), pigs (Esbenshade and Britt 1985), cattle (Johnson et al. 1988), and horses (Garza et al. 1986). Natural reversal occurs in the majority of immunized animals following a decline in antibody titers (Keeling and Crighton 1984). In captive white-tailed deer, a GnRH analogue conjugated to ovalbumin suppressed ovulation in only two of four females, even though antibody titers were measured in all four treated animals (Becker et al. 1999). Curtis and colleagues (2002) treated captive white-tailed deer with a GnRH immunocontraceptive and found that immunized females produced significantly fewer fawns than controls and had fewer estrous cycles than did either control or PZP-treated females. However, none of the vaccine-based methods is 100 percent effective because there is individual variation in immunocompetence.

METHODS FOR MALES

Steroid Hormones

PROGESTINS Twenty males of six ungulate species have been implanted with MGA, but all were treated in hopes of reducing aggression, not to effect contraception. Species included scimitar-horned oryx (*Oryx dammah*), addax (*Addax nasomaculatus*), nilgai (*Boselaphus tragocamelus*), gaur (*Bos frontalis*), Reeve's muntjac, and chital or spotted deer (*Axis axis*). Data on reproductive parameters were not collected, but the muntjac and nilgai males sired young during treatment, indicating that the dose was not contraceptive. Antler growth was reported to be abnormal or retarded in the chital and muntjac, and changes in coat color and horn growth were noted in the nilgai.

Three mule deer (*O. hemionus*) were treated with Depo-Provera, but females housed with them conceived, demonstrating inefficacy, at least at the doses used. Two male white-tailed deer and one camel (*Camelus* sp.) were each given one Depo-Provera injection but were subsequently lost to follow-up.

MGA administered to male cattle at doses that suppress reproduction in females has been reported not to affect male fertility (Lauderdale 1983), reproductive behavior, scrotal circumference, or overall circulating testosterone concentrations (Imwalle et al. 2002). However, total sperm production was decreased in adult male muntjac that consumed 0.3 to 1.0 mg MGA per animal per day, but a 6-month-old male still reached sexual maturity despite consuming MGA-treated feed (Stover et al. 1987). Male barasingha in a group of ungulates fed MGA (the dose consumed could not be measured) exhibited abnormal antler growth (Raphael et al. 1992). Antler growth in male fallow deer (*Dama dama*) treated for 7 months (May to January) with oral MGA (0.3 to 0.5 mg/animal/day) was twice the rate of untreated males (Wilson et al. 2002).

Protein Hormones

In contrast to data gathered from most other taxa (Jöchle, in preparation), evidence suggests that GnRH agonists are likely to be ineffective as contraceptives for male ungulates. Divergent responses in the reproductive–endocrine axis in males may result from differential effects of GnRH agonists on pituitary–gonadal function across species, which may occur in both males and females (Bergfeld et al. 1996). Several studies have revealed that treatment with GnRH agonists fails to sufficiently downregulate serum luteinizing hormone (LH) and testosterone to inhibit spermatogenesis (Melson et al. 1986; Apsden et al. 1997; Brinsko et al. 1998; D'Occhio et al. 2000; Penfold et al. 2002). This same pattern has been demonstrated in deslorelin-treated dorcas gazelle (*Gazella dorcas*), gerenuk (*Litocranius walleri walleri*), and scimitar-horned oryx (Penfold et al. 2002). In contrast, GnRH infusion of the agonist buserelin in male red deer (*Cervus elephus*) stimulated testicular growth and testosterone secretion and increased aggression (Lincoln 1987). In captive wild ungulates, neither leuprolide (Lupron) nor [D-tryptophen^{6}]luteinizing hormone-releasing hormone (D-Trp^{6}LH-RH) had any effects on testosterone levels or spermatogenesis in equids, including zebra and

Przewalski horses (*Equus przewalski*) (M. Briggs, personal communication). Finally, a male gaur treated four times with Lupron Depot (4-month depot injection; three injections: 90 mg/treatment) at 4-month intervals successfully sired offspring during and after treatment (M. Patton, unpublished data).

Eleven males of four species treated with GnRH agonists have been reported to the Contraception Database (see Table 9.1). Lupron Depot was administered to 1 male camel (11 injections), 1 giraffe (2 injections), and 3 bighorn sheep (1 injection each), but no further information was provided regarding success or even the reason for treatment, that is, aggression mitigation or contraception. In addition, 6 scimitar-horned oryx were treated with an unspecified GnRH agonist, but no follow-up information was provided.

Mechanical Methods

Silastic plugs injected into the vasa deferentia were reported for three fallow deer and one Gunther's dik dik (*Madoqua guentheri*). Three other ungulate species were part of a larger trial of vas plugs: impala, Himalayan tahr, and dama gazelle. However, when the plugs fit securely enough to successfully block semen, the procedure was not reversible (L. Zaneveld and C. Asa, unpublished data).

SUMMARY

Although most ungulates appear to experience few side effects from progestin-based contraceptives, occasional accumulation of fluid or mucus in the uterine lumen and uterine infection have been reported and merit continued monitoring. Progestins are not recommended for use in pregnant animals because of the possibility of prolonged gestation, stillbirth, and abortion, as reported for some species. When MGA is fed to mixed-sex groups of deer, antler growth and development may be abnormal. Currently, PZP vaccine appears to be a safe option for contraception of ungulates, although long-term and possibly nonreversible effects on the ovary need to be investigated more thoroughly. Also, PZP-treated animals may continue to exhibit estrous cycles beyond the typical breeding season, which may result in stress and, ultimately, in health problems for the adults or in young being born out of season. Also, adjuvants used to elicit an immune response, as well as delivery methods for vaccines, are not yet optimal. At present, the safest contraceptives for female ungulates appear to be GnRH agonists in implants or depot preparations, which seem to be without deleterious side effects. However, in males these agonists, in contrast to observations in other species, do not down-regulate LH sufficiently to suppress testicular function.

Timing of injections or implant administration for continued contraception is one of the most critical components in any contraceptive regimen, and failure to adhere to recommended schedules is the biggest cause of contraceptive failure. In contrast, apparent failure of contraceptive reversal in some individuals is something that is poorly understood and warrants further investigation.

ACKNOWLEDGMENTS

The authors thank Dr. Steve Monfort for his thorough editing of this chapter, as well as Dr. Jay Kirkpatrick and Kimberly Frank for information concerning porcine zona pellucida immunization.

REFERENCES

Apsden, W. J., A. Rao, P. T. Scott, I. J. Clarke, T. E. Trigg, J. Walsh, and M. J. D'Occhio. 1997. Differential responses in anterior pituitary luteinizing hormone (LH) content and LH α- and β-subunit mRNA, and plasma concentrations of LH and testosterone, in bulls treated with the LH releasing hormone agonist deslorelin. *Domest Anim Endocrinol* 14:429–437.

Baker, D. L., M. A. Wild, M. M. Conner, H, B. Ravivarapu, R. L. Dunn, and T. M. Nett. 2002. Effects of GnRH agonist (leuprolide) on reproduction and behavior in female wapiti (*Cervus elapus nelsoni*). *Reproduction Suppl* 60:155–167.

Becker, S. E., W. J. Enright, and L. S. Katz. 1999. Active immunization against gonadotropin-releasing hormone in female white-tailed deer. *Zoo Biol* 18:385–396.

Bergfeld, E. G., M. J. D'Occhio, and J. E. Kinder. 1996. Pituitary function, ovarian follicular growth, and plasma concentrations of 17 beta-estradiol and progesterone in prepubertal heifers during and after treatment with the luteinizing hormone-releasing hormone agonist deslorelin. *Biol Reprod* 4:776–782.

Brinsko, S. P., E. L. Squires, B. W. Pickett, and T. M. Nett. 1998. Gonadal and pituitary responsiveness of stallions is not down-regulated by prolonged pulsatile administration of GnRH. *J Androl* 19:100–109.

Brussow, K. P., F. Schneider, and W. Kanitz. 1996. Application of GnRH agonists in young sows affects cyclic gonadotropin release. *Tieraerztl Prax* 3:248–255.

Clarke, I. J., H. M. Fraser, and A. S. McNeilly. 1979. Active immunization of ewes against luteinizing hormone-releasing hormone, and its effects on ovulation and gonadotropin-releasing hormone neutralization. *Endocrinology* 110:1116–1123.

Curtis, P. D., R. L. Pooler, M. E. Richmond, L. A. Miller, G. F. Mattfeld, and F. W. Quimbly. 2002. Comparative effects of GnRH and porcine zona pellucida (PZP) immunocontraceptive vaccines for controlling reproduction in white-tailed deer (*Odocoileus virginianus*). *Reproduction Suppl* 60:131–141.

Deigert, F. A., A. E. Duncan, K. M. Frank, R. O. Lyda, and J. F. Kirkpatrick. 2003. Immunocontraception of captive exotic species. III. Contraception and population management of fallow deer (*Cervus dama*). *Zoo Biol* 22:261–268.

Delsink, A., J. J. Van Altena, J. F. Kirkpatrick, D. Grobler, and A. Fayrer-Hosken. 2002. Field applications of immunocontraception in African elephants (*Loxodonta africana*). *Reproduction Suppl* 60:117–124.

DeNicola, A. J., D. J. Kesler, and R. K. Swihart. 1997. Dose determination and efficacy of remotely delivered norgestomet implants on contraception of white-tailed deer. *Zoo Biol* 6:31–37.

D'Occhio, M. J., and W. J. Aspden. 1999. Endocrine and reproductive responses of male and female cattle to agonists of gonadotropin-releasing hormone. *J Reprod Fertil Suppl* 54:101–114.

D'Occhio, M. J., W. J. Aspden, and T. R. Whyte. 1996. Controlled reversible suppression of estrous cycles in beef heifers and cows using agonists of gonadotropin-releasing hormone. *J Anim Sci* 74:218–225.

D'Occhio, M. J., G. Fordyce, T. R. Whyte, W. J. Aspden, and T. E. Trigg. 2000. Reproductive responses of cattle to GnRH agonists. *Anim Reprod Sci* 60-61:433–442.

D'Occhio, M. J., G. Fordyce, T. R. Whyte, T. F. Jubb, L. A. Fitzpatrick, N. J. Cooper, W. J. Aspden, M. J. Bolam, and T. E. Trigg. 2002. Use of GnRH agonist implants for long-term suppression of fertility in extensively managed heifers and cows. *Anim Reprod Sci* 75:151–162.

Dunbar, B. S., S. Prasad, C. Carino, and S. M. Skinner. 2001. The ovary as an immune target. *J Soc Gynecol Invest* 8 (1 Suppl Proc): S43–S48.

Esbenshade, K. L., and J. H. Britt. 1985. Active immunization of gilts against gonadotropin-releasing hormone: effects on secretion of gonadotropins, reproductive function, and responses to agonists of gonadotropin-releasing hormone. *Biol Reprod* 333:569–577.

Frank, K. M., and J. F. Kirkpatrick. 2002. Porcine zona pellucida immunocontraceptive in captive exotic species: species differences, adjuvant protocols, and technical problems. In *Proceedings, American Association of Zoo Veterinarians*, 221–223, October 5–10, Milwaukee, WI.

Garza, F. Jr., D. L. Thompson Jr., D. D. French, J. J. Wiest, R. L. St. George, K. B. Ashley, L. S. Jones, P. S. Mitchell, and D. R. McNeill. 1986. Active immunization of intact mares against gonadotropin-releasing hormone: differential effects on secretion of luteinizing hormone and follicle-stimulating hormone. *Biol Reprod* 2:347–352.

Graham, L. H., T. Webster, M. Richards, K. Reid, and S. Joseph. 2002. Ovarian function in the Nile hippopotamus and the effects of depo-provera administration. *Reprod Fertil Suppl* 60:65–70.

Hayssen, V., A. van Tienhoven, and A. van Tienhoven. 1993. *Asdell's Patterns of Mammalian Reproduction: A Compendium of Species-Specific Data*. Ithaca: Comstock.

Herschler, R. C., and B. D. Vickery. 1981. Effects of (D-Tryp6-Des-Gly10-Pro-NH29)-luteinizing hormone-releasing hormone ethylamid on the estrous cycle, weight gain and feed efficiency in feedlot heifers. *Am J Vet Res* 42:1405–1408.

Imwalle, D. B., A. Daxenberger, and K. K. Schillo. 2002. Effects of melengesterol actetate on reproductive behavior and concentrations of LH and testosterone in bulls. *J Anim Sci* 80:1059–1067.

Jacobsen, N. K., D. A. Jessup, and D. J. Kesler. 1995. Contraception in captive black-tailed deer by remotely delivered norgestomet ballistic implants. *Wildl Soc Bull* 23:718–722.

Johnson, H. E., D. M. DeAvila, C. F. Chang, and J. J. Reeves. 1988. Active immunization of heifers against luteinizing hormone releasing hormone, human chorionic gonadotropin hormone and bovine luteinizing hormone. *Anim Sci* 66:719–726.

Keeling, B. J., and D. B. Crighton. 1984. Reversibility of the effects of active immunization against LHRH. In *Immunological Aspects of Reproduction in Mammals*, ed. D. E. Creighton, 370–397. London: Butterworths.

Kirkpatrick, J. F., P. P. Calle, P. Kalk, I. K. M. Liu, and J. W. Turner. 1996. Immunocontraception of captive exotic species. II. Formosan sika deer (*Cervus nippon taiouwanus*), axis deer (*Cervus axis*), Himalayan tahr (*Hemitragus jemlahicus*), Roosevelt elk (*Cervus elaphus roosevelti*), Reeve's muntjac (*Muntiacus reevesi*) and sambar deer (*Cervus unicolor*). *J Zoo Wildl Med* 27:482–495.

Lauderdale, J. W. L. 1983. Use of MGA (melengesterol acetate) in animal production. In *Anabolics in Animal Production*, ed. E. Meissonnier and J. Mitchell-Vigneron, 193–212. Paris: Office International des Epizooties.

Leuthold, W. 1977. *African Ungulates: A Comparative Review of Their Ethology and Behavioral Ecology*. New York: Springer.

Lincoln, G. A. 1987. Long-term stimulatory effects of a continuous infusion of LHRH agonist on testicular function in male red deer (*Cervus elaphus*). *J Reprod Fertil* 80:257–261.

McNeilly, A. S., and H. M. Fraser. 1987. Effect of gonadotropin-releasing hormone agonist-induced suppression of LH and FSH on follicle growth and corpus luteum function in the ewe. *J Endocrinol* 115:273–282.

McShea, W. J., S. L. Monfort, S. Hakim, J. F. Kirkpatrick, I. K. M. Liu, J. W. Turner, L. Chassy, and L. Munson. 1997. The effect of immunocontraception on the behavior and reproduction of white-tailed deer. *J Wildl Manag* 61:560–569.

Melson, B. E., J. L. Brown, H. M. Schoenemann, G. K. Tarnavsky, and J. J. Reeves. 1986. Elevation of serum testosterone during chronic LHRH agonist treatment in the bull. *J Anim Sci* 62:199–207.

Munson, L. 1993. Adverse effects of contraceptives in carnivores, primates and ungulates. In *Proceedings, American Association of Zoo Veterinarians*, 284–288, October 10–15, 1993, St. Louis, MO.

Munson, L. 2002. Reproductive diseases resulting from contraceptive treatments. In *Proceedings of the 2nd International Symposium on ART for the Conservation & Genetic Management of Wildlife*, 143–150, September 28–29, Omaha, NE.

Patton, M. L., M. S. Aubrey, M. Edwards, R. Rieches, J. Zuba, and V. Lance. 2000. Successful contraception in a herd of Chinese goral (*Nemorhaedus goral arnouxianus*). *J Zoo Wildl Med* 31:228–230.

Penfold, L. M., R. Ball, I. Burden, W. Jöchle, S. B. Citino, S. L. Monfort, and N. Wielebnowski. 2002. Case studies in antelope aggression control using a GnRH agonist. *Zoo Biol* 21:435–448.

Plotka, E. D., and U. S. Seal. 1989. Fertility control in female white-tailed deer. *J Wildl Dis* 25:643–646.

Raphael, B. L., P. P. Calle, W. B. Karesh, and R. A. Cook. 1992. Contraceptive program at the New York Zoological Society Institutions. In *Proceedings, Joint Meeting American Association of Zoo Veterinarians and the American Association of Wildlife Veterinarians*, 102–103, November 15–19, 1992, Oakland, CA.

Smith, D. E., M. E. Obrien, V. J. Palmer, and J. A. Sadowski. 1992. The selection of an adjuvant emulsion for a polyclonal antibody production using a low molecular weight antigen in rabbits. *Lab Anim Sci* 42:599–601.

Stover, J., R. Warren, and P. Kalk. 1987. Effect of melengesterol acetate on male muntjac (*Muntiacus reevesi*). In *Proceedings, First International Conference Zoological and Avian Medicine*, 387–388, September 6–11, 1987, Oahu, HI.

Tart, M. S. 1999. Fertility control in captive white-tailed does via GnRH agonist or prostaglandin F2α. Master of Science thesis. Athens, GA: University of Georgia.

Turner, A., and J. F. Kirkpatrick. 2002. Effects of immunocontraception on population, longevity and body condition in wild mares (*Equus caballus*). *Reproduction Suppl* 60:187–195.

Wilson, T. W., D. A. Neuendorff, A. W. Lewis, and R. D. Randel. 2002. Effect of zeranol or melengesterol acetate (MGA) on testicular and antler development and aggression in farmed fallow bucks. *J Anim Sci* 80:1433–1441.

Zimbelman, R. G., J. W. Lauderdale, J. H. Sokolowski, and T. G. Schalk. 1970. Safety and pharmacologic evaluations of melengestrol acetate in cattle and other animals: a review. *J Am Vet Med Assoc* 157:1528–1536.

PAUL P. CALLE

10
CONTRACEPTION IN PINNIPEDS AND CETACEANS

There are limited reports of contraceptive use in pinnipeds and cetaceans. Most of those that exist are based on anecdotal information, report a limited number of contraceptive bouts, or have not included a rigorous scientific experimental design (DeMatteo 1997; Robeck et al. 2001). Additionally, there is no information for these species on the effects of contraceptives on either pregnancy or lactation, and data (in the AZA Contraception Database) are insufficient to quantify efficacy or reversibility. Because there are few pinnipeds and cetaceans maintained in zoos and aquaria when compared to other nondomestic species, a greater emphasis has been placed upon the development of successful captive reproductive programs, rather than contraceptive techniques, for these marine mammals. However, effective population management often requires selective breeding, and the development of contraception techniques provides an additional management tool. Thus far, a few contraceptive techniques have been utilized, are generally recognized as effective, and are recommended for certain management conditions (Table 10.1) (Robeck et al. 2001; AZA CAG 2004).

Most pinnipeds have a highly seasonal reproductive pattern, in contrast to most cetaceans that exhibit estrous cycles year round, although some such as beluga whales (*Delphinapterus leucas*) exhibit strict reproductive seasonality. For seasonally breeding species, contraception can be achieved in some cases through limited intervention at the onset of, or only during, the reproductive season. Strategic intermittent administration can also be helpful in minimizing the potential for adverse effects that might be associated with contraceptive use.

Table 10.1
Contraceptive methods, number of individuals, and number of treatments for cetaceans and pinnipeds[a]

Method	Number of delivered treatments	Number of ongoing treatments	Number of species	*n*
Steroid hormones:				
Progestin implants:				
MGA implant	13	2	4	10
Progestin injections:				
Depo-Provera	76	—	2	12
Oral progestins:				
Ovaban	2	—	1	2
Supprestral	5	—	1	4♀/1♂
Regu-mate[b]	7	4	2	7
Protein hormones:				
Lupron Depot	38	—	3	8♂
Immunocontraception:				
PZP	38	—	2	12

MGA, melengestrol acetate; PZP, porcine zona pellucida.

[a]Data from the Contraception Database.

[b]Used to synchronize ovulation.

METHODS FOR FEMALES

Steroid Hormones

PROGESTINS The most common hormonal contraceptive techniques for female pinnipeds and cetaceans are administration of progestins through oral, injectable, or implant preparations (see Table 10.1; Robeck et al. 2001). The techniques reported include melengestrol acetate (MGA) implants (supplied by E. D. Plotka); oral megestrol acetate (Ovaban: Schering-Plough), medroxyprogesterone acetate (MPA, Provera: Pharmacia and Upjohn), or altrenogest (Regu-mate: Hoechst-Roussel); and injectable medroxyprogesterone acetate (MPA, Depo-Provera: Pharmacia and Upjohn).

PROGESTINS: MELENGESTROL ACETATE IMPLANTS Melengestrol acetate (MGA) implants have only been used in pinniped species, including the harbor seal (*Phoca vitulina*, $n = 4$); Northern fur seal (*Callorhinus ursinus*, $n = 3$); gray seal (*Halichoerus grypus*, $n = 3$); and California sea lion (*Zalophus californianus*, $n = 3$). Eight of the 11 completed bouts were 2 years or longer; in 5 cases the implants were never removed (the bout ended when the female died or was removed from the collection). No failures were reported.

PROGESTINS: INJECTIONS Medroxyprogesterone acetate (MPA: Depo-Provera) administered intramuscularly (IM) monthly at 200 mg (approximately 2 mg/kg) for 3 months during the breeding season (June through August) in female California sea lions was effective and reversible (Calle et al. 1993). The Contraception Database reveals one failure at the 200-mg dose, but whether this was the result of incomplete administration is not known. That female received five additional treatments during pregnancy, but no problems with parturition or with the infant were reported. Similar single doses (400 mg, approximately 2 mg/kg, IM) administered to female gray seals once at the onset of the breeding season (January) did not result in a high contraceptive rate, possibly because of suboptimal timing of the injection or dose for this species (Seely and Ronald 1991). Higher doses (600 mg, approximately 3 mg/kg, IM) administered monthly to female gray seals for 4 months through the breeding season (November through February) resulted in successful contraception at another location (L. Dunn, personal communication). Female harbor seals were also successfully treated at the same institution with similar doses (300 mg, approximately 3 mg/kg, IM monthly) administered through the breeding season (May through August) (L. Dunn, personal communication). However, MPA treatment resulted in a contraceptive failure in a female bottlenose dolphin (*Tursiops truncatus*; Dougherty et al. 2000).

PROGESTINS: ORAL The Contraception Database shows successful contraception was achieved in two gray seals treated with megestrol acetate (Ovaban, 60 mg daily) and four females (as well as the male in the group) treated with MPA (Supprestral, 100 mg weekly). The most common oral progestin used in female cetaceans has been altrenogest (Table 10.1; Robeck et al. 2001). Daily administration has been used for estrous synchronization in bottlenose dolphins (approximately 0.044 mg/kg body weight: Menchaca et al. 2002; 0.065 mg/kg: Robeck et al. 2000), Pacific white-sided dolphins (*Lagenorhynchus obliquidens*; approximately 0.044 mg/kg: Menchaca et al. 2002), and killer whales (*Orcinus orca*; 0.025 mg/kg: Robeck et al. 2000). Altrenogest treatment resulted in successful contraception in female bottlenose dolphins (approximately 0.044 mg/kg: Menchaca et al. 2002) and a female killer whale (0.044 mg/kg: Young and Huff 1996, 1997). Hormonal monitoring of urinary pregnanediol glucuronide and estrone conjugates, and of serum progesterone and estradiol 17β, in a treated killer whale demonstrated ovarian suppression (Young and Huff 1996, 1997). Successful reversal after 2 years of contraception (8.8 mg, approximately 0.04 mg/kg) has been documented in a female bottlenose dolphin (Dougherty et al. 2000; M. Menchaca, personal communication). Daily oral medroxyprogesterone acetate (5 mg, approximately 0.02 mg/kg) has been used successfully as a contraceptive for more than 10 years in a female bottlenose dolphin (N. van Elk, personal communication).

Immunocontraception

Porcine zona pellucida vaccine (PZP) has been administered to captive California sea lions, gray seals, harp seals (*Phoca groenlandica*), and hooded seals (*Cystophora cristata*) (Brown et al. 1997b; Renner and Dougherty 1999; Contraception Database), and free-ranging gray seals (Brown et al. 1996, 1997a; see Kirkpatrick and Frank, Chapter 13, Contraception in Free-Ranging Wildlife, this volume), for either contraceptive trials or determination of the immunological response to vaccination. These studies documented either elevated titers to antisoluble isolated zona pellucida antigen or successful contraception. PZP was also reported to be an effective contraceptive in free-ranging gray seals (Brown et al. 1996, 1997a). However, there were suspicions that adverse health consequences were attributable to use in captive California sea lions (M. Menchaca, personal communication). Because of the limited numbers of species and individuals treated and the inability to thoroughly follow up with free-ranging seals, conclusive information about efficacy, duration of action, or reversibility of PZP vaccination in pinnipeds is limited. It is therefore difficult to either recommend or caution its general use as a reversible contraceptive in pinnipeds. Preliminary data, however, suggest that vaccination with liposomal formulations of PZP may result in permanent contraception, compared to reversible contraception resulting from vaccination with native PZP (Brown et al. 1996; J. Kirkpatrick, personal communication). Immunocontraception has not been reported in cetaceans.

Surgical

Ovariohysterectomy and tubal ligation are not preferred techniques for permanent contraception in female pinnipeds or cetaceans because of the greater potential complications and challenges of both anesthesia and surgery in marine mammals for these more-involved, invasive procedures (Dierauf and Gulland 2001).

METHODS FOR MALES

Protein Hormones

The most common nonsteroidal contraceptives used in male pinnipeds and cetaceans are the gonadotropin-releasing hormone (GnRH) agonists depot leuprolide acetate (Depot Lupron: TAP Pharmaceuticals) and [D-tryptophen^6]luteinizing hormone-releasing hormone (D-Trp^{6}LH-RH) (Decapeptyl-CR: Ferring Pharmaceuticals) (see Table 10.1; Robeck et al. 2001). Administration of GnRH agonists results in an initial increase of follicle-stimulating hormone (FSH) and luteinizing hormone (LH) with concomitant increase in circulating testosterone levels, followed by testosterone suppression and decline to the levels of a castrated animal. These agents have been administered to male pinnipeds and cetaceans either for contraception or to suppress aggression in males during the breeding season.

Leuprolide acetate in a 1-month (0.11 to 0.19 mg/kg IM) or 4-month (0.9 to 1.1 mg/kg) depot preparation has been successfully administered to captive male California sea lions for reversible contraception (M. Briggs, personal communication). Single doses (7.5 mg, approximately 0.06 mg/kg IM) administered to male harbor seals in May, before the onset of the breeding season, resulted in testosterone suppression and successful contraception, with no change in aggressive or sexual behaviors (S. Hunter and B. Whitaker, personal communication). This treatment regimen resulted in successful contraception for 6 years in this colony. Rare side effects that have been observed in pinnipeds following leuprolide administration include anorexia, lethargy, and injection site discomfort and lameness (Calle et al. 1997).

Administration of leuprolide acetate to cetaceans has been limited to male bottlenose dolphins (see Table 10.1) (Briggs et al. 1995, 1996; Robeck et al. 2001). Monthly treatment (0.075 mg/kg IM) for months to years results in a reversible decrease in circulating testosterone levels, testicular atrophy, decreased sperm counts, and reversible contraception (Briggs et al. 1995, 1996; M. Briggs, personal communication).

Administration of a single dose (0.03 to 0.06 mg/kg IM) of D-Trp ^{6}-LH-RH resulted in testosterone suppression in both captive (Atkinson et al. 1993) and free-ranging (Atkinson et al. 1998) male Hawaiian monk seals (*Monachus schauinslandi*), with the captive seals demonstrating reversible suppression (Atkinson et al. 1993). Although contraceptive effect was not evaluated in these studies, the magnitude of testosterone decrease should have resulted in infertility. Administration also resulted in reversible testosterone suppression and decreased sexual and aggressive behaviors in captive male harbor seals (Yochem and Atkinson 1991; Atkinson and Yochem 1993). Infertility was observed after treatment of captive male harbor and gray seals administered D-Trp ^{6}-LH-RH (L. Dunn, personal communication).

Buserelin subcutaneous implants (9.9 mg, approximately 0.076 mg/kg each dose) (Suprefact-3-month depot; Aventis Pharma) administered twice in 14 days (April and May) to a captive male harbor seal resulted in reversible testosterone suppression and a decrease in sexual behaviors, expected to result in contraception (Siebert et al. 2002).

Surgical

Pinnipeds can be challenging anesthetic candidates, but with experience and care general anesthesia and surgical procedures may be safely performed. Minimizing anesthesia and surgical time results in fewer anesthetic complications. Because of the greater anesthetic risk and challenges of general anesthesia in cetaceans and the complications of wound healing in these totally aquatic species, general anesthesia and surgery are rarely conducted in cetaceans (Dierauf and Gulland 2001).

Castration is commonly preformed to achieve permanent contraception of male pinnipeds, with vasectomies less commonly employed (see Appendix; Robeck et al. 2001). Cetaceans have intraabdominal testes and are more difficult to safely anesthetize, so these sterilization techniques have not been developed for these species. Development of laparoscopic surgical techniques for cetaceans may provide an improved option for castration (Dover 2000).

SUMMARY

The recommendation for reversible contraception of female pinnipeds is injectable progestin (medroxyprogesterone acetate, Depo-Provera) administration through the breeding season (AZA CAG 2004). Oral progestin treatment may also be effective, but drug, dose, and efficacy have not been determined. Based upon information from other carnivores, in which progestin use can cause mammary gland carcinoma, endometrial hyperplasia, endometrial mineralization, endometritis, pyometra, or hydrometra, caution should be exercised and long-term use avoided (AZA CAG 2004; see Chapter 5, Adverse Effects of Contraceptives). The recommended reversible contraceptive for male pinnipeds is leuprolide acetate (Lupron Depot) before and throughout the breeding season (AZA CAG 2004). Because some treated pinnipeds experience injection site reactions (Calle et al. 1997), treated individuals should be monitored for adverse reactions and, if these occur, use should be discontinued. PZP vaccination of female pinnipeds can be considered if the potential for permanent contraception is acceptable, because the degree of reversibility is uncertain at this time (Robeck et al. 2001; AZA CAG 2004; see Chapter 13, Contraception in Free-Ranging Wildlife). Permanent male contraception can be achieved through either castration or vasectomy (Robeck et al. 2001; AZA CAG 2004).

Deslorelin, a new GnRH agonist implant (Suprelorin), holds promise for extended reversible gonadal suppression that may be utilized for contraception of male or female pinnipeds or as an agent to diminish aggression in male pinnipeds during the breeding season (Calle et al. 1999; Bertschinger et al. 2001). Other GnRH agonists such as D-Trp ^{6}-LH-RH or buserelin may also be effectively utilized (Robeck et al. 2001; Siebert et al. 2002).

The recommendation for reversible contraception of female cetaceans is oral altrenogest (Robeck et al. 2001; AZA CAG 2004). This drug has been used in a variety of cetacean species for both estrous synchronization and contraception without adverse effects being reported. Although other oral progestins may also be effective, they have not been used extensively. Recommendation for reversible male contraception is the administration of leuprolide acetate (Lupron Depot: Robeck et al. 2001; AZA CAG 2004). Although this method requires repeated IM injections, adverse reactions have not been reported in cetaceans.

Marine mammals reproduce frequently in many zoos and aquaria, and the success of these propagation programs has fostered an increasing need to regulate the reproduction of these animals to manage genetic diversity in the population. Contraception is one tool that is being used to meet this need. The same techniques are also sometimes necessary for treatment of medical conditions, or to stop reproductive activity for health reasons, or are utilized in assisted reproduction efforts. The techniques outlined in this chapter represent the initial efforts by those involved in marine mammal reproduction programs to achieve these goals. Future work will expand upon these efforts, identify limitations of current programs, and develop new and better contraceptive techniques for marine mammals.

REFERENCES

American Zoo and Aquarium Association Contraception Advisory Group (AZA CAG). *Annual Recommendations 2004.* www.stlzoo.org/contraception.

Atkinson, S., and P. K. Yochem. 1993. The use of GnRH agonists for fertility and behavior control in pinnipeds. In *Symposium: Contraception in Wildlife Management*, 6–7, October 26–28, 1993, Denver, CO.

Atkinson, S., W. G. Gilmartin, and B. L. Lasley. 1993. Testosterone response to a gonadotrophin releasing hormone agonist in Hawaiian monk seals (*Monachus schauinslandi*). *J Reprod Fertil* 97:35–38.

Atkinson, S., T. J. Ragen, W. G. Gilmartin, B. L. Becker, and T. C. Johanos. 1998. Use of a GnRH agonist to suppress testosterone in wild male Hawaiian monk seals (*Monachus schauinslandi*). *Gen Comp Endocrinol* 112:178–182.

Bertschinger, H. J., C. S. Asa, P. P. Calle, J. A. Long, K. Bauman, K. DeMatteo, W. Jochle, T. E. Trigg, and A. Human. 2001. Control of reproduction and sex related behaviour in exotic wild carnivores with the GnRH analogue deslorelin: preliminary observations. *J Reprod Fertil Suppl* 57:275–283.

Briggs, M. B., W. Van Bonn, R. M. Linnehan, D. Messinger, C. Messinger, and S. Ridgway. 1995. Effects of leuprolide acetate in depot suspension on testosterone levels, testicular size, and semen production in male Atlantic bottlenose dolphins (*Tursiops truncatus*). *Proc Int Assoc Aquat Anim Med* 26:112–114.

Briggs, M. B., D. Messinger, C. Messinger, R. M. Linnehan, W. Van Bonn, S. Ridgway, and G. Miller. 1996. Effects of leuprolide acetate in depot suspension on the testosterone levels, testicular size, and semen production in male Atlantic bottlenose dolphins (*Tursiops truncatus*). In *Proceedings of the Annual Conference of the American Association of Zoo Veterinarians*, 330–333, November 3–8, 1996, Puerto Vallarta, Mexico.

Brown, R. G., W. C. Kimmins, M. Mezei, J. Parsons, and B. Pohajdak. 1996. Birth control for grey seals. *Nature* (Lond) 379:30–31.

Brown, R. G., W. D. Bowen, J. D. Eddington, W. C. Kimmins, M. Mezei, J. L. Parsons, and B. Pohajdak. 1997a. Evidence for a long-lasting single administration contraceptive vaccine in wild grey seals. *J Reprod Immunol* 35:43–51.

Brown, R. G., W. D. Bowen, J. D. Eddington, W. C. Kimmins, M. Mezei, J. L. Parsons, and B. Pohajdak. 1997b. Temporal trends in antibody production in captive grey, harp and hooded seals to a single administration immunocontraceptive vaccine. *J Reprod Immunol* 35:53–64.

Calle, P. P., B. L. Raphael, and R. A. Cook. 1993. Contraception of California sea lions (*Zalophus californianus*) with injectable medroxyprogesterone acetate. *Proc Int Assoc Aquat Anim Med* 24:46.

Calle, P. P., M. D. Stetter, B. L. Raphael, R. A. Cook, C. McClave, J. Basinger, H. Walters, and K. Walsh. 1997. Use of depo leuprolide acetate to control undesirable male associated behaviors in the California sea lion (*Zalophus californianus*) and California sea otter (*Enhydra lutris*). *Proc Int Assoc Aquat Anim Med* 28:6–7.

Calle, P. P., B. L. Raphael, R. A. Cook, C. McClave, J. A. Basinger, and H. Walters. 1999. Use of depot leuprolide, cyproterone, and deslorelin to control aggression in an all male California sea otter *(Enhydra lutris nereis)* colony. *Proc Int Assoc Aquat Anim Med* 30:42–45.

DeMatteo, K. 1997. *AZA Contraception Advisory Group Contraception Report. III. Carnivores and Other Mammals*. St. Louis: St. Louis Zoological Park.

Dierauf, L. A., and F. M. D. Gulland, ed. 2001. *CRC Handbook of Marine Mammal Medicine*. Boca Raton: CRC Press.

Dougherty, M., G. Bossart, and M. Renner. 2000. Case study: calf born thirteen months post-regumate contraception with social complications. In *Bottlenose Dolphin Reproduc-*

tion Workshop, ed. D. Duffield and T. Robeck, June 3–6, 1999, 41–42. Silver Springs, MD: AZA Marine Mammal Taxon Advisory Group.

Dover, S. 2000. Laparoscopic techniques for the bottlenose dolphin (*Tursiops truncatus*) and potential for reproductive management. In *Bottlenose Dolphin Reproduction Workshop*, ed. D. Duffield and T. Robeck, June 3–6, 1999, 205–206. Silver Springs, MD: AZA Marine Mammal Taxon Advisory Group.

Menchaca, M. M., R. Rose, H. Gorman, and S. Graff. 2002. Regu-mate® in reproductive management: a synchronizing project using Regu-mate® in six cetaceans resulted in five ovulations at a specific time with two subsequent pregnancies. *Proc Int Assoc Aquat Anim Med* 33:94–96.

Renner, M. S., and M. Dougherty. 1999. Preliminary investigation of the safety and efficacy of porcine zona pellucida (PZP) immunocontraception in captive California sea lions (*Zalophus californianus*). *Proc Int Assoc Aquat Anim Med* 30:91.

Robeck, T. R., E. Jensen, F. Brook, N. Rourke, C. Rayner, and R. Kinoshita. 2000. Preliminary investigations into ovulation manipulation techniques in delphinids. In *Proceedings, American Association of Zoo Veterinarians/International Association of Aquatic Animal Medicine Joint Conference*, 222–225, September 17–21, 2000, New Orleans, LA.

Robeck, T.R., S. K. C. Atkinson, and F. Brook. 2001. Reproduction. In *CRC Handbook of Marine Mammal Medicine*, ed. L. A. Dierauf and F. M. D. Gulland, 193–236. Boca Raton: CRC Press.

Seely, A. J,. and K. Ronald. 1991. The effect of depo-provera on reproduction in the grey seal (*Halichoerus grypus*). In *Proceedings of the Annual Meeting of the American Association of Zoo Veterinarians*, 304–310, September 28–October 3, 1991, Calgary, Alberta, Canada.

Siebert, U., J. Driver, T. Rosenberger, and J. Sandow. 2002. Reproduction control in a group of common seals (*Phoca vitulina*) through the use of a gonadotropin-releasing hormone (GnRH Suprefact®-3-Month Depot). *Proc Int Assoc Aquat Anim Med* 33:111.

Young, J. F., and D. G. Huff. 1996. Fertility management in a female killer whale (*Orcinus orca*) with altrenogest (Regu-mate®). *Proc Int Assoc Aquat Anim Med* 27:66.

Young, J. F., and D. G. Huff. 1997. Animal breeding management: synchronizing estrus in a killer whale (*Orcinus orca*) with altrenogest (Regu-mate®). *Proc Int Assoc Aquat Anim Med* 28:63.

Yochem, P. K., and S. Atkinson. 1991. Effects of GnRH agonist on plasma testosterone concentrations and socio-sexual behavior of harbor seals. In *9th Biennial Conference on the Biology of Marine Mammals*, 3, December 5–9, 1991, Chicago, IL.

KAREN E. DEMATTEO

11
CONTRACEPTION IN OTHER MAMMALS

Carnivores, primates, and even some ungulate species maintained in zoos have a longer history of contraceptive use than "other mammals" such as marsupials and bats, likely because cooperative captive breeding programs for these taxa are relatively recent. This chapter summarizes the data reported to the American Zoo and Aquarium Association (AZA) Contraception Advisory Group (CAG) through December 2001 on the types of reversible contraception used in 6 mammalian orders for which we have information: Hyracoidea, Scandentia, Rodentia, Edentata, Chiroptera, and Marsupialia. A review of the Appendix shows that, although the Contraception Database contains information for 25 species of "other mammals," more than half (54 percent) are represented by only 1 female whereas only 2 species are represented by 10 or more individuals. Six reversible contraceptive methods have been used in small mammals, and four of the six are progestin based. Of the total 117 contraceptive treatments, 87 (74 percent) are the melengestrol acetate (MGA) implant (supplied by E. D. Plotka).

Obviously, a meaningful evaluation for any of the data presented is premature at best. Instead, we provide this review to illustrate the limited scope of the data currently available and to provide a general summary. The methods used to collect information for the Contraception Database are presented in Part III: The Application.

METHODS FOR FEMALES

Steroid Hormones

PROGESTINS: MELENGESTROL ACETATE IMPLANTS

Order Hyracoidea: Six MGA implants have been used in five female rock hyrax. Duration of use ranged from 10 to 102 months; there were no contraceptive failures (Table 11.1).

Order Scandentia: The only representative of this order (see Appendix) is a single female tree shrew (*Tupaia glis*) treated with an MGA implant. The treatment is ongoing.

Order Edentata: The only representatives of this order are one female from each of two species (see Appendix) treated with an MGA implant. A tamandua (*Tamandua tetradactyla*) was effectively treated for 3 months, and a Hoffman's sloth (*Choloepus hoffmanni*) contraceptive bout is ongoing.

Order Rodentia: Within this order, 17 MGA implants have been used in 15 females from eight species. One implant, in a female springhare (*Pedetes capensis*), was lost. Capybaras (*Hydrochaeris hydrochaeris*) account for more than half the implants used in rodents. Of the 9 females treated, 1 was already pregnant (Table 11.2), and 3 of the remaining 8 females conceived with the implant known to be in place. The high failure rate in capybara may be species specific, and therefore a higher dose must be tested in this taxon.

Order Chiroptera: The Rodrigues fruit bat (*Pteropus rodricensis*), the only bat species for which there are data, has only been treated with MGA implants (see Table 11.1). Twenty of the 48 MGA implant bouts represent completed contraceptive treatments; the remaining 28 are ongoing (Table 11.1). Four implants were lost, 4 females were pregnant when implanted (see Table 11.2), and 1 failure occurred. MGA implants were found to be 93 percent effective in Chiroptera (based on implants in use for 2 years; $n = 15$). Hayes et al. (1996) determined there were no behavioral effects of MGA implants in the highly social Rodrigues fruit bat when compared to nonimplanted individuals. However, implanted individuals had weight gain, retarded hair growth at the incision site, and a 22 percent implant loss rate. Despite these problems, Hayes et al. (1996) concluded that MGA implants were an effective contraceptive for this species.

Table 11.1
Contraceptive methods, number of individuals, and number of treatments for other mammals[a]

Method	Order	Number of delivered treatments	Number of ongoing treatments	Number of species	*n*
Steroid hormones:					
Progestin implants:					
MGA	Hyracoidea	6	0	1	5♀
	Scandentia	1	1	1	1♀
	Rodentia	17	13	8	15♀
	Endentata	2	1	2	2♀
	Chiroptera	48	28	1	31♀
	Marsupialia	13	5	6	11♀
Norplant I	Rodentia	3	0	2	3♀
	Rodentia	7	0	1	7♀
	Marsupialia	10	0	6	9♀/1♂
Progestin injections:					
Proligestone	Rodentia	5	—	1	2♀
Depo-Provera	Marsupialia	2	—	1	1♀
Protein hormones:					
GnRH agonist:					
Deslorelin	Rodentia	2	0	1	1♂
Mechanical:					
Vas plugs	Marsupialia	1	0	1	1♂

MGA, melengestrol acetate; GnRH, gonadotropin-releasing hormone.

[a]Data from the Contraception Database.

Order Marsupialia: Six species of marsupials have been treated with MGA implants. Of the 13 MGA implants, 6 were used in one species, the wallaroo (*Macropus robustus*), and no contraceptive failures have been reported. The implant of one female wallaroo, treated for 28 months, was removed to allow reproduction, but she had not conceived after 12 months.

PROGESTINS: OTHER IMPLANTS

Order Rodentia: Three rodent species have been treated with levonorgestrel implants (Norplant; Wyeth): a single beaver (*Castor*

Table 11.2
Successful parturition in other mammals implanted with MGA during pregnancy[a]

Order	Species	n	Trimester contraception started	Trimester contraception stopped	Successful parturition (n)
Steroid hormones:					
Progestin implants:					
MGA implant:					
Rodentia	*Hydrochaeris hydrochaeris*	1	2	—	1
Chiroptera	*Pteropus rodricencis*	2	1	—	2
		2	2	—	2
Norplant II					
Rodentia	*Dolichotis patagonum*	1	2	—	1

MGA, melengestrol acetate.

[a]Data from the Contraception Database.

canadensis), two Malagasy giant jumping rats (*Hypogeomys antimena*), and seven mara (*Dolichotis patagonum*). Varied doses and durations of use further complicate data interpretation for such a small sample size. One female mara was already pregnant when implanted with Norplant (see Table 11.2), and no failures were reported.

Order Marsupialia: Five marsupial species, including a female black wallaroo (*Macropus bernardus*, n = 1), western grey kangaroos (*M. fuliginosus*, n = 4), an eastern grey kangaroo (*M. giganteus*, n = 1), red-legged pademelons (*Thylogale stigmatica*, n = 2), and a swamp wallaby (*Wallabia bicolor*, n = 1) (see Table 11.1), were treated with Norplant implants at doses that ranged from 2.5 to 21 mg/kg body weight. Much of the variation resulted from the recommendation that the Norplant implant not be cut; thus, the entire implant was used in different-sized animals. No failures were reported.

In studies of tammar wallabies (*M. eugenii*) and eastern grey kangaroos, Nave et al. (2000, 2002) determined that levonorgestrel (LNG) implants were an effective and reversible long-term contraceptive technique. In both species, LNG did not prevent reactivation or subsequent development of embryos in diapause that were conceived before initiation of treatment. Births occurred normally, and lactation and development of young appeared unaffected. Fertility resumed after implants were removed, indicating that LNG can be an effective and reversible contraceptive in these species.

PROGESTINS: INJECTIONS

Order Marsupialia: Only two medroxyprogesterone acetate injections (Depo-Provera: Pharmacia and Upjohn) were administered to a single marsupial female (see Table 11.1). However, because the injections were not given consecutively, the doses varied, and the second injection was given when an MGA implant was still in place, efficacy cannot be properly evaluated.

Order Rodentia: Proligestone, a second-generation progestin not produced in the United States, was the only injectable steroidal contraceptive reported in Rodentia (Table 11.1). There were few repetitions, and the problem of interpretation is compounded because dosages were not provided.

Immunocontraception

No reports on immunocontraceptive use have been submitted to the Contraception Database. However, an antisperm vaccine has been tested in brush-tailed possums (*Trichosurus vulpecula*; Duckworth et al. 1998). Subsequent to injections of 5×10^7 sperm in Freund's complete adjuvant followed by boosters with Freund's incomplete adjuvant, females had a significant reduction in fertility. These results suggest that possum sperm contains antigens that could function as immunocontraceptives.

METHODS FOR MALES

Steroid Hormones

PROGESTINS

Order Marsupialia: A single LNG implant (Norplant II; Wyeth) was used in a male yellow-footed rock wallaby (*Petrogale xanthopus*) at a dose of 6.7 mg/kg body weight in an effort to reduce aggression, which proved successful; however, fertility was not assessed.

Protein Hormones

Order Rodentia: The GnRH agonist implant deslorelin (Suprelorin: Peptech Animal Health) was used in male mara. Because there are only two completed treatments, however, the technique cannot be properly evaluated.

Mechanical Methods

Order Marsupialia: Vas deferens plugs were tested in one red kangaroo (*Macropus rufus*). Although sperm passage was blocked, the procedure was not reversible, providing no advantage over vasectomy (L. Zaneveld and C. Asa, unpublished data).

Surgical Methods

Notwithstanding that these taxon-based chapters cover only reversible contraceptive methods, an interesting study of brush-tailed possums (*Trichosurus vulpecula*) by Jolly et al. (1999) provides information on the effects of sterilizing dominant males to limit reproduction. The authors examined the use of vasectomy in social groups containing multiple males, and the vasectomized, dominant males did indeed control access to fertile females.

SUMMARY

The application of contraceptive methods to control reproduction in this diverse group of "other mammals" is in its infancy, and a considerable amount of research is needed to determine which methods are effective, reversible, and safe for these species. To date, the predominant methods of contraception that have been used are progestins, primarily MGA implants but also Norplant. The scant information available indicates that progestins are effective, which is most evident for fruit bats, the taxon with the largest sample size. It also may be true for capybara, if the cause of the MGA implant failures is dose- and not method related. However, research should not be limited to steroid hormones but rather should also be directed at other promising methods such as deslorelin, currently being tested in primates and carnivores. Indeed, all mammalian taxa benefit from having a variety of contraceptive options that can be used to address particular social or husbandry conditions.

ACKNOWLEDGMENTS

This summary would not have been possible without the time and effort invested by all the institutions that participated in the research program and completed annual CAG surveys. The technical assistance of Bess Frank in some of the preliminary Contraception Database summaries was greatly appreciated. The comments provided by anonymous reviewers on earlier drafts of this chapter were valued.

REFERENCES

Duckworth, J. A., B. M. Buddle, and S. Scobie. 1998. Fertility of brushtail possum (*Trichosurus vulpecula*) immunized against sperm. *J Reprod Immunol* 37:125–138.

Hayes, K. R., A. T. C. Feistner, and E. C. Halliwell. 1996. The effect of contraceptive implants on the behavior of female Rodrigues fruit bats, *Pteropus rodricensis*. *Zoo Biol* 15:21–36.

Jolly, S. E., E. B. Spurr, and P. E. Cowan. 1999. Social dominance and breeding success in captive brushtail possums, *Trichosurus vulpecula*. *N Z J Zool* 26:21–25.

Nave, C. D., G. Shaw, R. V. Short, and M. B. Renfree. 2000. Contraceptive effects of levonorgestrel implants in marsupials. *Reprod Fertil Dev* 12:81–86.

Nave, C. D., G. Coulson, A. Poiani, G. Shaw, and M. B. Renfree. 2002. Fertility control in the eastern grey kangaroo using levonorgestrel implants. *J Wildl Manag* 66:470–477.

LINDA M. PENFOLD, MARILYN L. PATTON,
AND WOLFGANG JÖCHLE

12
CONTRACEPTIVE AGENTS IN AGGRESSION CONTROL

This chapter does not attempt to examine the extent and role of aggressive behavior or to cover all aspects of aggression control. Instead, as there is a growing amount of information on the usefulness of contraceptive agents as moderators of aggressive behavior, our aim is to discuss the "side effects" of contraceptives as potential agents for aggression control. Because the male of the species is usually more aggressive than the female, only male aggression is covered in this chapter.

Aggression in many species is primarily an adaptation to cope with competition for resources such as mate access, territory, or food (Moynihan 1998). Aggression may manifest itself overtly, in the form of physical confrontation, or may be subtle, involving the use of displays and songs. Although there are obvious advantages to behaviors that result in better survival or reproductive success, aggressive behaviors have the disadvantage in that they are costly in terms of energy and time and may result in injury or death (Moynihan 1998). Among captive animals, limited space may precipitate more overt aggressive behaviors, which must then be managed to prevent serious injury or death. Careful management may prevent or minimize aggressive interactions using strategies such as food placement in multiple areas in enclosures or removal of problem animals (Judge et al. 1994), but even when these and other management options are available, physiological methods for addressing aggression may still be needed.

In males, testosterone has been shown to play an important role in aggressive behavior, and it has been well documented that castration of domestic animals can reduce aggression (Martin and Lindsay 1998; Knol and Egberink-Alink 1989).

Thus, the use of certain contraceptives that reduce testosterone concentrations may be useful for aggression control. Castration is the most effective approach to reduce androgen concentrations, but because aggressive behavior is not always correlated with circulating androgens (Logan and Wingfield 1990), the effects of castration vary greatly. In historic times men were gonadectomized (eunuchs) for a variety of social or religious reasons, but these castrations resulted in a variety of responses depending upon the age when the testes were removed and whether the penis was also amputated. Castration of prisoners of war, as practiced over millennia, often reduced neither their sexual prowess nor their fighting spirit (Jöchle 2001). Regardless of the consequences, the effects of castration are irreversible and thus it is an unacceptable option for many zoo animals. In this chapter, we consider how "chemical castration" may provide a more reasonable, as well as reversible, method of aggression control.

UNGULATES

In zoos and wildlife institutions, maintaining surplus male ungulates is often a challenge. It is well known that male ungulates are more aggressive than females. Hence, there is a rich folklore of combat between male camels as well as rams (Bouissou 1983). In captivity, limited by space, excess males are usually housed as a bachelor group, but crowding alone can exacerbate aggressive behaviors. Male groups may often be successful for a period of time during which the group is stable, but social relationships may break down when a new animal is introduced, where there is competition for resources, or as the males age. Fighting behaviors can also increase during the breeding season, even in the absence of females.

Even though aggressive behaviors were found to be independent of circulating testosterone in bulls (Price et al. 1986) and male lambs (Ruiz-de-la-Torre and Manteca 1999), aggressive behavior has been reported as androgen dependent in other ungulates (Bouissou 1983). Therefore, reducing circulating androgen concentrations could be an effective way to reduce aggression. However, evidence of that theory is inconsistent and, in some cases, contradictory (see following).

Progestins

Because aggression control has become such an important part of male ungulate captive management, progestins have been tried in several studies to decrease agonistic behaviors. Progestins act as antagonists to androgens (Knol and Egberink-Alink 1989), but the dose given may affect results. For example, in one study, the progestin melengestrol acetate (MGA), when fed to domestic bulls, abolished surges of testosterone soon after treatment (Haynes et al. 1977), and bulls receiving supplemental progesterone exhibited disrupted spermatogenesis, indicating inhibited testosterone secretion (Matsuyama et al. 1967). In another study, however, 1.0 and 2.0 mg MGA per day per animal did not significantly reduce serum testosterone in peripubertal, sexually naive bulls (Imwalle et al. 2002). Progestins also have been used with exotic hoofstock. For example, a subjective decrease in aggressive behavior was noted in muntjacs (*Muntiacus reevesi*; Stover et al. 1987) after oral administration of 0.2 mg/day MGA in feed for 60 days and in scimitar-horned oryx (*Oryx dammah*; Blumer et al. 1992) treated with MGA implants for 6 months. Orally administered MGA (2.77 mg/day/animal) also reduced aggression and fecal androgen levels in a bachelor herd of fringe-eared oryx (*Oryx gazella callotis*; Patton et al. 2001). Paradoxically, when MGA was fed to male fallow deer (*Dama dama*), reduced aggression, concurrent with increased testosterone concentrations, was noted in the treated bucks (Wilson et al. 2002). In this case, it is possible that androgens other than testosterone, such as dihydrotestosterone, are more responsible for regulating aggression in this species. However, in contrast to the previous studies, where progestin treatment did modify aggressive behavior, five gerenuk (*Litocranius walleri walleri*) treated with the progestin medoxyprogesterone acetate (MPA, Depo-Provera: Pharmacia and Upjohn) for 2 months followed by an MGA implant failed to show any decrease in aggressive behavior over a 7-month period, even though serum testosterone concentrations were dramatically reduced (Penfold et al. 2002).

There are several theories as to why progestins appear to modify behavior in some ungulate species but not in others, even though testosterone is reduced. One possibility is that the mechanism of action of progestins in aggression control is mediated less by reduction in circulating testosterone concentrations and more by an anxiolytic effect. Another explanation may be that the adrenal androgen dehydroepiandrosterone (DHEA), a steroid hormone relatively unaffected by contraceptive agents that target the hypothalamic-pituitary-gonadal axis, may be the primary androgen mediating aggressive behavior.

Gonadotropin-Releasing Hormone Agonists

The mechanism of action of gonadotropin-releasing hormone (GnRH) agonists in bovid species has proven to be different than that of other species (see Chapter 3, Types of Contraception, and Chapter 9, Contraception in Ungulates). In ungulates, treatments with GnRH and their analogues prevent pulsatile luteinizing hormone (LH) release but not tonic LH production, which may actually increase over time. Consequently, testosterone remains unchanged and may rise with time. Studies have shown that GnRH agonists do not suppress testosterone concentrations in the domestic bull (D'Occhio and Aspden 1996; D'Occhio et al. 2000), domestic ram (Jimenez-Severiano et al. 1999), donkey, stallion (Brinsko et al. 1998), gerenuk, dorcas gazelle, and scimitar-horned oryx (Penfold et al. 2002). No detectable changes in aggressive behavior were seen in gerenuk and scimitar-horned oryx treated with deslorelin implants (Suprelorin: Peptech Animal Health), and aggressive behavior was also unchanged in dorcas gazelle treated with deslorelin or leuprolide acetate (Lupron Depot: TAP Pharmaceuticals; Penfold et al. 2002). Similarly, treatment with another GnRH agonist, buserelin, failed to reduce LH and circulating testosterone in red deer stags (*Cervus elaphus*). In that study, continuous infusion of buserelin resulted in increased LH and testosterone secretion that in turn actually resulted in a significant increase in aggressive behavior (Lincoln 1987).

Antigonadotropin-Releasing Hormone Vaccine

Immunization against GnRH is the most promising alternative to castration for reducing male aggressive behavior in adult animals (D'Occhio 1993). Because the molecular structure of GnRH is conserved between mammalian species (Sherwood et al. 1993), immunization against it suppresses release of gonadotropins from the pituitary, resulting in reduced (nondetectable) LH and follicle-stimulating hormone (FSH) concentrations as well as reduced gonadal function in all species. A major area of interest in using this technique involves elimination of unwanted male aggressiveness (Thompson 2000). Reports of decreased aggression in immunocastrated animals exist for bulls (Thompson 2000) and goat bucks (Godfrey et al. 1996). In mature stallions, the effect of immunization varied between stallions, but the vaccine was effective in suppressing testicular function and androgen secretion (Malmgren et al. 2001) and could prove to be a mechanism for controlling aggression in equids.

PRIMATES

Zoological institutions have an interest in keeping both multimale mixed-sex groups and all-male groups of primates for social and management reasons. However, multimale and bachelor groups are often problematic, and as juvenile males reach adulthood, intermale aggression may become serious. Hormonal control of male aggressive behaviors may allow some primates to be maintained together. Studies investigating the relationship between testosterone and aggression in some macaque species have provided mixed results, however. In primates, the interactions among androgens, aggression, dominance rank, and sexual activity are quite complex and context dependent (Dixson 1980).

Progestins

Cyproterone acetate (CA), a synthetic steroid, is referred to as an antiandrogen. Studies indicate CA has strong progestin and weak androgen agonist activity (Eaton et al. 1999). In a group of 34 adult male rhesus monkeys (*Macaca mulatta*), aggression was found to be dependent on androgen levels (Rose et al. 1971), and both aggressive behavior and testosterone were reduced when treated with CA. However, the authors questioned the relative contributions of the addition of progestin and decline in testosterone on aggression and suggested an alternative explanation, that the modified aggression was independent of testosterone and progestins may act directly on the neural centers involved in aggression (Eaton et al. 1999).

Male long-tailed macaques (*Macaca fascicularis*) treated with MPA had decreased testosterone levels, increased aggressive behavior toward other males, but decreased aggression toward females. In this case, reduced testosterone modified aggressive behavior that males directed toward females within a sexual context but did not modify intermale aggression (Zumpe et al. 1991). In contrast, male stump-tailed macaques (*Macaca arctoides*) treated with MPA displayed increased aggres-sion toward subordinate females, lending support to the idea that MPA may have aggression-enhancing effects on certain types of aggressive behavior (Linn and Steklis 1990). In rats, MPA has been shown to be androgenic (Labrie et al. 1987), possibly explaining these increases in aggression in some macaques. On the other hand, male chimpanzees (*Pan troglodytes*) treated with MPA were found to have reduced aggression and serum testosterone levels (Orkin 1993).

Gonadotropin-Releasing Hormone Agonists

Gonadotropin-releasing hormone agonists have been shown to be effective in modifying aggression in lion-tailed macaques (*Macaca silenus*: Norton et al. 2000; I. Porton, personal communication). Two males previously displaying aggressive behavior toward one another were successfully housed together, with no aggressive behavior for more than 4 years, after a single treatment with 12 mg deslorelin. As the effects of the implant wore off, aggressive behavior slowly returned over a period of several months, starting as minor dominance behaviors such as displacement from food, before obvious aggression was observed (Norton et al. 2000). In another case, four adult male lion-tailed macaques were first treated with a histrelin implant for 2 years, which was replaced with a 12-mg dose of deslorelin. The males have lived compatibly in an all-male group for more than 7 years (I. Porton, personal communication). In black lemurs (*Eulemur macaco*) and ringtailed lemurs (*Lemur catta*), repeated use of the GnRH agonist deslorelin

(3 mg) reduced testosterone blood concentrations and testis size significantly (C. Williams, personal communication, 1999–2000), and some male black lemurs changed pelage temporarily from male to female phenotype. The effect on aggression, measured by whether males had to be separated from conspecifics because of unacceptable levels of aggression, was variable, and additional research is recommended (C. Williams and I. Porton, personal communication).

CARNIVORES

Progestins

Although progestins have been shown to be effective in reducing aggressive behavior toward conspecifics in the domestic dog (W. Jöchle, unpublished data), they do not always modify aggressive behavior of the dogs toward their owners (Hart 1991). Thus, in a zoo setting, control of aggression in canids may be possible for conspecifics, but perhaps not for aggression manifested toward keeper staff. Progestins have also reduced aggression in male lions, but in this instance the mechanism of action was entirely different, because progestin therapy administered to females to prevent estrous behavior was sufficient to prevent aggressive behavior in the male (W. Jöchle, unpublished data). Thus, it is likely that any treatment suppressing estrus in the female may have the desired effect of reducing aggression in the male. However, the use of progestins is contraindicated in both canids and felids because of adverse side effects such as the induction of mammary gland tumors, genital tract pathology in females, precipitation of diabetes mellitus in prediabetic patients, and adrenal suppression (see Chapter 5, Adverse Effects of Contraceptives). Thus, only extreme cases would warrant treatment using this method.

Gonadotropin-Releasing Hormone Agonists

It seems that GnRH analogues are not only effective contraceptives for carnivore species (see Chapter 7, Contraception in Carnivores) but also are safe and effective for aggression control. In a black-footed cat (*Felis nigripes*), a single deslorelin implant (6 mg) significantly reduced aggression, testis size, and blood testosterone (Bertschinger et al. 2001) for several months. In 18 male cheetahs (*Acinonyx jubatus*), deslorelin rendered both sperm and testosterone undetectable for almost 2 years (Bertschinger et al. 2002), although aggressive behavior was not monitored. Likewise, GnRH agonists, particularly deslorelin implants, were very effective in temporarily reducing aggression in male domestic dogs and male African wild dogs (Trigg et al. 2001; Bertschinger et al. 2001, 2002). In the California sea otter

(*Enhydra lutris*), treatment with GnRH agonists (Lupron or deslorelin implants) eliminated male intraspecies aggression so long as the animals were biologically neutered (Bertschinger et al. 2001). Another GnRH agonist, [D-tryptophen[6]] luteinizing hormone-releasing hormone (D-Trp[6]LH-RH) (Decapeptyl-CR: Ferring Pharmaceuticals), suppressed testosterone in wild male Hawaiian monk seals (*Monachus schauinslandi*), and although aggressive behavior was not monitored in this species, it is likely that it may reduce mortality in the monk seal caused by adult male aggression related to reproduction and mating behavior (Atkinson et al. 1998).

SUMMARY

There is considerable variation in the effect on aggression among different species treated with hormones or their analogues. These variations could be due to either one or a combination of treatment effects, such as dose, duration of treatment, season, age, species, and demography of housing, or, in the case of GnRH agonists used in ungulates, a difference in the mechanism of biological action.

There is mounting evidence that the behavior-modifying effects of various progestin contraceptives such as MGA and MPA may be independent of circulating testosterone concentrations. Reduced androgens are not the sole hormonal pathway mediating the effects of progestins on aggression, and studies have suggested alternative mechanisms. For example, in primates behavioral effects of MPA appear to be mediated by brain mechanisms regulating sexual motivation that are relatively independent of circulating androgen levels (Michael et al. 1991). Glucocorticoids have also been implicated in regulating aggression (Knol and Egberink-Alink 1989), and studies indicate that MGA acts first as a progestin and secondarily as a glucocorticoid (Lauderdale 1983). Finally, reduced aggression may be a consequence of a progestational "highly potent, naturally occurring hypnotic-anesthetic agent" effect (Gyermek 1967). In addition, aggressive patterns of behavior can be learned, become habit, and be expressed independent of androgen support.

Studies continue to reveal inconsistencies in the relationship between androgen levels and aggressive behavior. Aggressive behavior is complex and dependent on many external factors, such as competition for food or mates, and may be a manifestation of an innate behavior, explaining why it is so difficult to reliably and consistently control aggressive behavior. Studies suggesting that certain receptors or pathways may be "primed" by high testosterone concentrations before, or soon after, birth in males show that finding a broad solution to aggressive behavior in captive and free-ranging wildlife will be unlikely. Research continues to show the complex relationship between gonadal hormones and aggression. The hormonal or chemical control of aggression is poorly understood, and both successes and failures have accompanied the use of contraceptive agents. In spite of the variation in results, we suggest that the further study of contraceptive agents as a means of controlling aggression in captive animals will prove beneficial.

REFERENCES

Atkinson, S., T. J. Ragen, W. G. Gilmartin, B. L. Becker, and T. C. Johanos. 1998. Use of a GnRH agonist to suppress testosterone in wild male Hawaiian monk seals (*Monachus schauinslandi*). *Gen Comp Endocrinol* 112:178–182.

Bertschinger, H. J., C. S. Asa, P. P. Calle, J. A. Long, K. Bauman, K. DeMatteo, W. Jochle, T. E. Trigg, and A. Human. 2001. Control of reproduction and sex related behaviour in exotic wild carnivores with the GnRH analogue deslorelin: preliminary observations. *J Reprod Fertil Suppl* 57:275–283.

Bertschinger, H. J., T. E. Trigg, W. Jöchle, and A. Human. 2002. Induction of contraception in some African wild carnivores by downregulation of LH and FSH secretion using the GnRH analogue deslorelin. *Reproduction Suppl* 60:41–52.

Blumer, E. S., E. D. Plotka, and W. B. Foxworth. 1992. Hormonal implants to control aggression in bachelor herds of scimitar horned oryx (*Oryx dammah*): a progress report. In *Proceedings of the Joint Conference of the American Association of Zoo Veterinarians and the American Association of Wildlife Veterinarians*, 212–216, November 15–19, 1992, Oakland, CA.

Bouissou, M. F. 1983. Androgens, aggressive behaviour and social relationships in higher mammals. *Horm Res* (Basel) 18:43–61.

Brinsko, S. P., E. L. Squires, B. W. Pickett, and T. M. Nett. 1998. Gonadal and pituitary responsiveness of stallions is not down-regulated by prolonged pulsatile administration of GnRH. *J Androl* 19:100–109.

Dixson, A. F. 1980. Androgens and aggressive behavior in primates: a review. *Aggress Behav* 6:37–67.

D'Occhio, M. J. 1993. Immunological suppression of reproductive functions in male and female mammals. *Anim Reprod Sci* 33:345–372.

D'Occhio, M. J., and W. J. Aspden. 1996. Characteristics of luteinizing hormone (LH) and testosterone secretion, pituitary responses to LH-releasing hormone (LHRH) and reproductive function in young bulls receiving the LHRH agonist deslorelin: effect of castration on LH responses to LHRH. *Biol Reprod* 54:45–52.

D'Occhio, M. J., G. Fordyce, T. R. Whyte, W. J. Aspden, and T. E. Trigg. 2000. Reproductive responses of cattle to GnRH agonists. *Anim Reprod Sci* 60/61:433–442.

Eaton, G. G., J. M. Worlein, S. T. Kelley, S. Vijayaraghavan, D. L. Hess, M. K. Axthelm, and C. L. Bethea. 1999. Self-injurious behavior is decreased by cyproterone acetate in adult male rhesus (*Macaca mulatta*). *Horm Behav* 35:195–203.

Godfrey, S. I., S. W. Walkden-Brown, G. B. Martin, and E. J. Speijers. 1996. Immunization of goat bucks against GnRH to prevent seasonal reproductive and agonistic behavior. *Anim Reprod Sci* 44:41–54.

Gyermek, L. 1967. Pregnanolone: a highly potent, naturally occurring hypnotic-anesthetic agent. *Proc Soc Exp Biol Med* 125:1058–1062.

Haynes, N. B., T. E. Kiser, H. D. Hafs, and J. D. Marks. 1977. Prostaglandin $F_{2\alpha}$ overcomes blockade of episodic LH secretion with testosterone, melengestrol acetate or aspirin in bulls. *Biol Reprod* 17:723–728.

Hart, B. L. 1981. Progestin therapy for aggressive behavior in male dogs. *J Am Vet Med Assoc* 178:1070–1071.

Imwalle, D B., A. Daxenberger, and K. K. Schillo. 2002. Effects of melengesterol acetate on reproductive behavior and concentrations of LH and testosterone in bulls. *J Anim Sci* 80:1059–1067.

Jimenez-Severiano, H., M. Mussard, T. Davis, W. D. Enwright, M. D'Occhio, and J. Kinder. 1999. Secretion of luteinizing hormone (LH) in rams chronically treated with analogs of gonadotropin-releasing hormone (GnRH). *Biol Reprod* (Suppl I) 60:276.

Jöchle, W. 2001. Hormonal approaches to modifying social aggression in captive wildlife. In *Proceedings, AZA Wildlife Contraception Center Symposium, Exploring Methods for Maintaining Social Groups by Reducing Aggression*, 17–34, September 5, 2001, St. Louis, MO.

Judge, P. G., B. M. de Waal, K. S. Paul, and T. P. Gordon. 1994. Removal of a trauma-inflicting alpha matriline from a group of rhesus macaques to control severe wounding. *Lab Anim Sci* 44:344–350.

Knol, B. W., and S. T. Egberink-Alink. 1989. Androgens, progestogens and agonistic behaviors, a review. *Vet Q* 11:94–101.

Labrie, C., L. Cusan, M. Plante, S. Lapointe, and F. Labrie. 1987. Analysis of the androgenic activity of synthetic "progestins" currently used for the treatment of prostate cancer. *J Steroid Biochem* 28:379–384.

Lauderdale, J. W. 1983. Use of MGA (melengestrol acetate) in animal production. In *Anabolics in Animal Production Symposium*, 193–212, February 15–17, 1983. OIE (World Organisation for Animal Health), Paris, France.

Lincoln, G. A. 1987. Long-term stimulatory effects of a continuous infusion of LHRH agonist on testicular function in male red deer (*Cervus elaphus*). *J Reprod Fertil* 80: 257–261.

Linn, G. S., and H. D. Steklis. 1990. The effect of depo-medroxyprogesterone acetate (DMPA) on copulation-related and agonistic behaviors in an island colony of stump-tailed macaques (*Macaca arctoides*). *Physiol Behav* 47:403–408.

Logan, C. A., and J. C. Wingfield. 1990. Autumnal territorial aggression is independent of plasma testosterone in mockingbirds. *Horm Behav* 24:568–581.

Malmgren, L., O. Andresen, and A. M. Dalin. 2001. Effect of GnRH immunization on hormonal levels, sexual behaviour, semen quality and testicular morphology in mature stallions. *Equine Vet J* 1:75–83.

Martin, G. B., and D. R. Lindsay. 1998. Castration, effects in nonhuman mammals (male). In *Encyclopedia of Reproduction*, ed. E. Knobil and J. D. Neill, 486–496. San Diego: Academic Press.

Matsuyama, S., M. Richkind, and P. T. Cupps. 1967. Effects of supplemental progesterone on semen from bulls. *J Dairy Sci* 50:375–377.

Michael, R. P., R. W. Bonsall, and D. Zumpe. 1991. Medroxyprogesterone acetate and the nuclear uptake of testosterone and its metabolites by brain, pituitary gland and genital tract in male cynomolgus monkeys. *J Steroid Biochem Mol Biol* 38:49–57.

Moynihan, M. H. 1998. *The Social Regulation of Competition and Aggression in Animals*. Washington, DC: Smithsonian Institution Press.

Munson, L., I. A. Gardner, R. J. Mason, L. M. Chassy, and U. S. Seal. 2002. Endometrial hyperplasia and mineralization in zoo felids treated with melengestrol acetate contraceptives. *Vet Pathol* 39:419–427.

Norton, T. M., L. M. Penfold, B. Lessnau, W. Jöchle, S. K. Staaden, A. Jolliffe, J. E. Bauman, and J. Spratt. 2000. Long acting deslorelin implants to control aggression in male lion-tailed macaques (*Macaca silenus*). In *Proceedings of the American Association of Zoo Veterinarians and the International Association of Aquatic Animal Medicine*, 174–177, September 17–21, New Orleans, LA.

Orkin, J. L. 1993. Use of medroxyprogesterone acetate to reduce aggression in male chimpanzee (*Pan troglodytes*). *Lab Anim Sci* 43:260–261.

Patton, M. L., A. M. White, R. R. Swaisgood, R. L. Sproul, G. A. Fetter, J. Kennedy, M. S. Edwards, R. G. Rieches, and V. A. Lance. 2001. Aggression control in a bachelor herd of fringe-eared oryx (*Oryx gazella callotis*) with melengesterol acetate: behavioral and endocrine observations. *Zoo Biol* 20:375–388.

Penfold, L. M., R. Ball, I. Burden, W. Jöchle, S. B. Citino, S. L. Monfort, and N. Wielebnowski. 2002. Case studies in antelope aggression control using a GnRH agonist. *Zoo Biol* 21:435–448.

Price, E. O., L. S. Katz, G. P. Moberg, and S. J. R. Wallach. 1986. Inability to predict sexual and aggressive behaviors in plasma concentrations of testosterone and luteinizing hormone in hereford bulls. *J Anim Sci* 42:613–17.

Rose, R. M., J. W. Holaday, and I. S. Bernstein. 1971. Plasma testosterone, dominance rank and aggression behaviour in male rhesus monkeys. *Nature* (Lond) 231:366–368.

Ruiz-de-la-Torre, R. L., and X. Manteca. 1999. Effects of testosterone on aggressive behavior after social mixing in male lambs. *Physiol Behav* 68:109–113.

Sherwood, N. H., D. A. Lovejoy, and I. R. Coe. 1993. Origin of mammalian gonadotropin-releasing hormones. *Endocr Rev* 14:241–254.

Stover, J., R. Warren, and P. Kalk. 1987. Effect of melengesterol acetate on male muntjac (*Muntiacus reevesi*). In *Proceedings of the 1st International Conference of Zoological Avian Medicine*, 387–388, Oahu, HI.

Thompson, D. L. Jr. 2000. Immunization against GnRH in male species (comparative aspects). *Anim Reprod Sci* 60/61:459–469.

Trigg, T. E., P. J. Wright, A. F. Armour, P. E. Williamson, A. Junaidi, G. B. Martin, A. G. Doyle, and J. Walsh. 2001. Use of a GnRH analogue implant to produce reversible long-term suppression of reproductive function in male and female domestic dogs. *J Reprod Fertil Suppl* 57:255–261.

Wilson, T. W., D. A. Neuendorff, A. W. Lewis, and R. D. Randel. 2002. Effect of zeranol and melengesterol acetate (MGA) on testicular and antler development and aggression in farmed fallow bucks. *J Anim Sci* 80:1433–1441.

Zumpe, D., R. W. Bonsall, M. H. Kutner, and R. P. Michael. 1991. Medroxyprogesterone acetate, aggression and sexual behavior in male cynomolgus monkeys (*Macaca fascicularis*). *Horm Behav* 25:394–409.

JAY F. KIRKPATRICK AND KIMBERLY M. FRANK

13
CONTRACEPTION IN FREE-RANGING WILDLIFE

In recent years, the concept of controlling free-ranging wildlife populations by nonlethal means, particularly through contraception, has received increasing attention. The reasons for this new emphasis on fertility control as a population management tool for free-ranging species are multiple and diverse. Some species are afforded protection by virtue of their location, in national parks or other legally designated refuges that do not permit hunting or other lethal methods such as poisoning or trapping, whereas other species, such as the North American wild horse (*Equus caballus*), have been protected through federal legislation (throughout this chapter the term wild, instead of feral, is used to identify North American free-ranging horses, to remain consistent with the Free-Roaming Wild Horse and Burro Act). White-tailed deer (*Odocoileus virginianus*) have found themselves in the protective custody of increased urbanization, in areas where hunting is no longer legal, wise, or safe, or sometimes not acceptable to the public. Finally, certain exotic species that are threatening to native flora and fauna must be controlled to protect fragile habitats and ecosystems.

In many instances, changing societal values and the resulting public opposition to lethal controls have also changed the direction of management philosophies. The general public clearly does not view wildlife management in the same manner it did 50 years ago (Herzog et al. 2001), and this trend is worldwide in scope. It is important here to make clear that these changing values are not just a function of the views and activities of animal protection groups or organized efforts to halt lethal controls, but represent a general philosophical shift in thinking by a larger public. Opposition to waterfowl hunting and the shooting of kangaroos is common in Australia, fox hunting has lost its popularity in Britain, and the culling of elephants has created a split in the attitudes of Africans.

At the same time, the application of contraception to captive exotic species has become a common and increasingly important management tool for zoo personnel and has been demonstrated to be a highly successful approach to controlling some animal populations (Seal 1991). The application of contraception to free-roaming wildlife as opposed to captive exotic species, however, introduces a variety of new problems and challenges to the field of wildlife fertility control. The first issue is that of contraceptive delivery where the animals are not confined, usually quite wary, and seldom disposed to facilitate the application of the contraceptive. This particular dimension of the problem brings form and substance to the "art" of wildlife contraception, in addition to the science of this arcane subdiscipline. A second challenge involves regulatory issues, because treated free-roaming wildlife may represent food animals for scavengers, predators, or even humans. A third issue is whether or not the particular wildlife species is perceived as "desirable" or a "pest" by the public, because that attitude in turn will dictate the specific parameters of the contraceptive, whether reversible or not reversible, that may be employed. These are only a few of the unique problems attendant to altering reproductive success in free-ranging species.

Before 1980, there are few published reports of attempts at contraception of free-ranging wildlife and even fewer successes. Since that time, an increasingly large research effort has been mounted to address the problems of wildlife fertility control, but even after more than 20 years the actual number of applications in the field, as opposed to research with controlled populations, is strikingly small and reflects the difficulties attendant on this endeavor. Thus, despite the large body of literature reflecting sound research with captive animals, this review focuses almost exclusively on published actual field applications, successful or unsuccessful. Despite the focus here on field applications, it is noteworthy that contraceptive research and application within the zoo community have made significant contributions to fertility control in free-ranging wildlife by providing captive models under controlled conditions before moving into the field. A number of studies of sterilization in various wildlife species are reported in the literature, but these are not included here because they do not, by definition, concern contraception.

A brief examination of the history of contraception in wild canids and cervids is instructive, if only to highlight some of the problems associated with actual application of fertility control in the field despite technology that showed promise in a captive setting. Several compounds were tested during the 1950s and 1960s that showed pharmacological promise for inhibiting fertility in coyotes (*Canis la-*

trans) (Balser 1964; Gates et al. 1976; Thompson 1976) and two species of foxes (Linhart 1963; Linhart and Enders 1964; Cheatum and Hansel 1967). However, when field tests were conducted, bait acceptance became a serious problem. Linhart (1964) found that bait acceptance by foxes was sporadic and that nontarget species consumed the baits. Oleyar and McGinnes (1974) discovered that gray foxes (*Urocyon cinereoargenteus*) would take baits but red foxes (*Vulpes vulpes*) would not. Although Balser (1964) showed that diethylstilbestrol (DES) would inhibit fertility in captive coyotes, when it was presented in tallow baits to wild animals it was not ingested with any regularity and was poorly absorbed from the fat-based bait (Brushman et al. 1967); thus, the reproductive season often was simply delayed.

In another study of oral delivery of DES to red foxes (Allen 1982), 50 mg DES and 125 mg tetracycline hydrochloride (TC) were incorporated into pork tallow baits coated with sugar and fed to free-ranging foxes over a 4-year period. The baits were placed once each year between February 1 and February 13. An average of 70 to 75 percent of foxes inhabiting the study area consumed at least one bait, based on TC fluorescence in mandibular bone, and there was no difference between male and female consumption. Despite the success in delivering the bait, there was no difference in fertility between control and treated groups. Thus, although captive studies showed promise, the reality of application to wild animals in the field yielded disappointing results. The only successful field study of contraception in canids, with African hunting dogs (*Lycaon pictus*), occurred recently (Bertschinger et al. 2002). In this study, the gonadotropin-releasing hormone (GnRH) antagonist deslorelin was placed in long-acting implants (Suprelorin: Peptech Animal Health) after capture.

Similar results emerged from studies with deer and elk. A number of steroid compounds were shown to be effective contraceptives in captive white-tailed deer (Harder 1971; Harder and Peterle 1974; Bell and Peterle 1975; Roughton 1979) and wapiti (*Cervus elaphus*; Greer et al. 1968). However, subsequent field tests showed that steroid implants were impractical (Levenson 1984), that bait acceptance was too unreliable in the case of oral delivery (Matschke 1977), and that often the drug had to be consumed on a daily basis to be effective (Roughton 1979). Once again, the difficulties of delivering contraceptive drugs to free-ranging wildlife negated the pharmacological successes with captive animals. The same history surrounds attempts at rodent contraception, and similar problems rendered contraception of free-roaming rodents ineffective.

Much research has been conducted in the search for new and effective contraceptive agents for wildlife (Kirkpatrick and Turner 1984, 1991; Seal 1991), but disappointingly few efforts have found their way to actual field tests or management-level applications. Unfortunately, many attempts at the development of wildlife contraceptives have occurred without much thought to the unique logistical and physiological problems associated with free-ranging animal populations. Until 1991, no attempts had even been made to hypothesize an "ideal" wildlife contraceptive. Thus, research was often aimed wide of the mark. Although some of the characteristics of such a contraceptive that emerged were arbitrary, there was some agreement that a pattern for success had been established. Before reviewing these characteristics, it is important to understand that these were constructed around the issue of North American wild horses; thus, there are some peculiarities that do not necessarily apply to other species.

The proposed ideal contraceptive for wild horses included the following characteristics: (1) contraceptive effectiveness of at least 90 percent; (2) safety when administered to pregnant females; (3) reversibility of contraceptive action; (4) relatively minor cost; (5) absence of short- or long-term health side effects; (6) ability to be delivered remotely, without handling animals; (7) minimal effects on individual and social behaviors; and (8) no passage through the food chain (Kirkpatrick and Turner 1991).

The issue of reversibility is not so important in some other species, such as the feral species inhabiting national parks, or species perceived as pests. In North American wild horses, however, public opinion and political pressures require that only reversible contraceptives be applied. The issue of safety in pregnant animals is less important in white-tailed deer, for example, where females are nonpregnant for half the year, as opposed to horses, which have an 11-month gestation that results in pregnancy almost year round. Also, behavioral effects are of lesser importance in species with less complex social organization. Thus, although the stated characteristics are general, each one has greater or lesser importance with any particular species.

At the same time, a fundamental three-step approach to wildlife contraception was envisioned (Kirkpatrick and Turner 1995). Step one consisted of answering the question of whether a particular agent was capable of inhibiting fertility in a particular species. Domestic animals and captive zoo animals provide a wonderful opportunity to answer this question without the expense and complicated logistics of working in the field. Step two was the question of whether the agent could actually be delivered to wild animals under field conditions. The importance of this question is immense. It is one thing to treat a captive deer and quite another to get a dart into a deer that has the ability to elude hunters with real guns. However, if the drug cannot be delivered, it has no practical value. Step three was the question of whether a population effect could be achieved in the field, which is the ultimate goal of any wildlife contraceptive effort. This last question is really just another way of asking if a sufficiently large portion of the population can be treated in any given year. These three steps are absolutely fundamental, but steps two and three usually escape the general public's grasp of the concept of wildlife contraception, and leave the would-be consumer with a concern only whether there is an agent that can actually inhibit fertility in a particular species.

This brief historical review should frame the difficulties of wildlife contraception beyond the pure science of contraceptive compounds. It is within this complex milieu that more recent and sometimes successful wildlife contraception has evolved. In actuality, there are additional constraints to successful wildlife contraception, in the form of political opposition, cultural objections and taboos, and even economic arguments that entwine themselves throughout the issue, but these are not addressed here. A summary of free-ranging species that have been treated with contraceptives is given in Table 13.1.

EQUIDS

In 1971, the issue of wildlife contraception was energized with the passage of the Free-Roaming Wild Horse and Burro Act, which gave almost total protection to horses and burros living on public lands in the United States. This legislation, however, was passed without any evidence of concern for future population control. During the ensuing 10 years, research explored inhibition of stallion fertility by administering a long-acting androgenic compound, testosterone propionate (TP). The theory was that the androgen would feed back to the hypothalamus and/or pituitary and block the release of luteinizing hormone (LH), thereby preventing spermatogenesis. After lengthy and successful trials with domestic pony stallions, Kirkpatrick et al. (1982) and Turner and Kirkpatrick (1982) reported successfully inhibiting reproduction in free-ranging wild stallions by 83 percent in Challis, ID, by lowering sperm counts and motility. Each stallion was given 2.5 to 10 g TP, incorporated in slow-release lactide-glycolide microcapsules (Southern Research Institute) suspended in polyethylene glycol. Although the approach worked pharmacologically, the animals had to be immobilized from a helicopter and then hand injected. The resulting stress and danger of injury rendered this approach impractical.

In a subsequent experiment designed to test the effectiveness of male contraceptives, wild stallions were vasectomized and reproductive success was studied in their respective bands (Asa 1999). Within the bands held by vasectomized stallions, 17 and 33 percent of bands (n = 40 in two areas) produced foals, indicating that either bachelor stallions or subordinate stallions were breeding mares successfully. Among the bands headed by intact stallions, 86 and 88 percent of bands had foals.

Table 13.1
Studies of contraceptives in free-ranging wildlife

Family	Species	Method	Reference
Equidae	*Equus caballus*	Testosterone proprionate	Kirkpatrick et al. 1982; Turner and Kirkpatrick 1982; Kirkpatrick 1995
		Vasectomy	Asa 1999
		Estrogen/progesterone	Plotka and Vevea 1990; Eagle et al. 1992
		GnRH vaccine	Goodloe et al. 1996
		PZP vaccine	Kirkpatrick et al. 1990, 1991, 1992, 1995; Turner and Kirkpatrick 2002; Kirkpatrick and Turner 2002; Turner et al. 1997, 2001, 2002; Cameron et al. 2001; Stafford et al. 2001
	Asinus asinus	PZP vaccine	Turner et al. 1996a
Cervidae	*Odocoileus virginianus*	MGA	Levenson 1984; Roughton 1979
		DES	Matschke 1977
		hCG vaccine	DeNicola et al. 1996
		Prostaglandin $F_{2\alpha}$	DeNicola et al. 1997; Becker and Katz 1994
		PZP vaccine	Turner et al. 1992, 1996b; McShea et al. 1997; Naugle et al. 2002
Canidae	*Canis latrans*	DES	Brushman et al. 1967
	Urocyon cinereoargenteus	DES	Oleyar and McGinnes 1974
	Vulpes fulva	DES	Allen 1982
	Lycaon pictus	Deslorelin	Bertschinger et al. 2002
Felidae	*Panthera leo*	MGA	Berry 1996; Orford 1978, 1996; Orford and Perrin 1988
	Felis catus	Megestrol acetate	Remfry 1978; McDonald 1980
	Panthera pardus	Delorelin	Bertschinger et al. 2001, 2002
	Acinonyx jubatus	Deslorelin	Bertschinger et al. 2002
Elephantidae	*Loxodonta africana*	PZP vaccine	Fayrer-Hosken et al. 1999, 2000; Nave et al. 2000
Petauridae	*Macropus eugenii*	Levonoregestrel	Nave et al. 2000
	Macropus gigantus	Levonorgestrel	Nave et al. 2001, 2002; Poiani et al. 2002
Phocidae	*Halichoerus grypus*	PZP vaccine	Brown et al. 1996, 1997a, 1997b
Aplondontidae	*Cyanomys ludovicanus*	DES	Garrott and Franklin 1983
Mustelida	*Mephitis mephitis*	DES	Storm and Sanderson 1969
		Levonorgestrel	Bickle et al. 1991
Procyonidae	*Procyon lotor*	Levonorgestrel	Kramer 1996

GnRH, gonadotropin-releasing hormone; PZP, porcine zona pellucida; MGA, melengestrol acetate; DES, diethylstilbestrol; hCG, human chorionic gonadotropin.

This work more or less ended attempts at stallion-focused fertility control in wild populations.

In a second attempt to inhibit fertility in wild horses, Silastic (Dow Corning) rods containing estradiol (E, 4 g), progesterone (P, 6 g), or ethinylestradiol (EE, 1.5 to 4 g) placed in captive and free-ranging mares reduced reproductive rates by 75 to 100 percent in the captive groups (Plotka and Vevea 1990; Eagle et al. 1992). In this approach, the steroids would interfere with the feedback mechanisms for LH, and possibly for follicle-stimulating hormone (FSH) as well, to prevent follicular development or ovulation. Today, however, we know that neither the TP-based approach with stallions nor the estrogen–progestin-based approach with mares would be permitted because of regulatory issues associated with passage of these compounds through the food chain.

By the 1980s, interest was expanding from steroids to immunocontraception. Goodloe et al. (1996) tested GnRH conjugated to keyhole limpet hemocyanin (KLH) in wild mares at Cumberland Island National Seashore, GA. In theory, the GnRH vaccine would prevent ovulation by blocking the action of LH. The vaccine was freeze-dried and delivered by biodegradable biobullets (BallistiVet) but, although significant antibodies were formed against the vaccine, contraception failed.

A breakthrough occurred in 1988, when, based on the earlier work of Liu et al. (1989) with captive horses, a porcine zona pellucida (PZP) vaccine was delivered remotely to 26 wild mares at Assateague Island National Seashore (ASIS). None of the mares that received an initial inoculation of 65 μg PZP plus Freund's complete adjuvant (FCA) followed by one or two 65-μg booster inoculations plus Freund's incomplete adjuvant (FIA), administered between February and April by means of darts, were pregnant in 1989. Furthermore, 15 of the mares that were pregnant at the time of inoculation delivered healthy foals (Kirkpatrick et al. 1990). The two-inoculation regimen worked as well as the three-shot regimen. Antibodies against PZP caused steric hindrance of the sperm receptor on the mare's own zona pellucida and blocked fertilization (Liu et al. 1989). A year later, half the animals were given a single booster inoculation with FIA, and only 1 of 14 produced a foal (Kirkpatrick et al. 1991). Of related interest was the fact that only two small (25-mm) abscesses occurred at injection sites among all treated animals. In the earlier study, several of the captive animals (Liu et al. 1989) that were inoculated in the neck formed serious abscesses, but in the ASIS studies all inoculations were given in the hip, in the gluteal or semitendinosus muscles.

In a related study, Turner et al. (1996a) demonstrated that the same methodology and dosages worked equally well in free-ranging burros (*Equus asinus asinus*) at Virgin Islands National Park. The significance of this work was to show that the same approach would work in nonseasonally breeding equids.

This initial work, and another 3 years of research on ASIS (Kirkpatrick et al. 1992, 1995), led to the first management-level application of contraception in wild horses or, for that matter, in any wildlife. In 1994 the National Park Service instituted management of the 166-animal herd on ASIS using the PZP vaccine. The outcome of that work was that zero population growth was attained in a single year, and the herd has not grown significantly since 1995 (Turner and Kirkpatrick 2002). Over that same period of time it was shown that body condition scores increased, mare and foal mortality decreased, and entirely new age classes appeared among these animals, extending the expected life span by nearly 10 years. Reversibility of contraceptive effects has been demonstrated in animals treated 1 (100 percent), 2 (100 percent), 3 (70 percent), 4 (100 percent), and 5 (100 percent) consecutive years, but a small group of mares (n = 5) treated for 7 consecutive years have not yet returned to fertility (Kirkpatrick et al. 1992, 1995; Kirkpatrick and Turner 2002) after a 9-year hiatus. This long-term research/management program has even demonstrated that mares whose mothers were treated while they were in utero were fertile when they reached sexual maturity. The total cost of this management is approximately $1,500 annually, in addition to the cost of labor. Normally, it takes two shooters about 2 weeks in March and one shooter for another week in August to treat the 50 to 70 mares annually, with a single booster inoculation.

The management-level application of PZP to ASIS mares also indicated that there was a simple alternative to giving two inoculations in the first year. There was some urgency in getting contraceptive management underway because of the rapid increase in the population; however, contraceptive treatment could not begin before a required environmental assessment was completed in 1995. In March 1994, all 73 mares that had never been previously treated were given a single 65-μg PZP plus FCA inoculation, with the intent of merely causing antigen recognition and establishing all animals as "one-shot" mares thereafter, in preparation for management in 1995. The outcome was a surprising 70 percent efficacy from the single inoculation (Turner and Kirkpatrick 2002), which probably resulted from the timing of the inoculation, just before the breeding season, and the use of a powerful adjuvant, FCA. In 1995, all those mares needed only a single 65-μg PZP plus FIA inoculation to maintain the contraceptive effects. The practical point of this experiment was that, if the managers are willing to give only a single inoculation the first year and forgo immediate contraceptive effects, the animals require only single annual booster inoculations thereafter.

Beyond contraceptive efficacy, both short- and long-term safety is a major consideration for managers of wild horses, which seem to attract a good deal of protective attention from wild horse advocacy groups. The 14 years of study on ASIS has provided valuable data regarding reversibility, safety in pregnant animals or animals in utero when the mother was inoculated, and long-term health of treated animals. Reversibility of contraceptive effects is certain among animals treated for 2 to 5 consecutive years, but the time for reversal is highly variable, ranging from 1 to 7 years. Beyond 5 consecutive years of treatment, there is no evidence for reversibility. During the 12 years of study, there has been no evidence of PZP immunocontraception extending the breeding season, nor has survival differed between foals born to treated or untreated mares (Kirkpatrick and Turner 2003). Finally, females that were carried in utero in treated mares were equally fertile, when attaining adulthood, to those born to untreated mothers (Kirkpatrick and Turner 2002).

The ASIS studies provided many valuable lessons on the importance of remote delivery systems (Kirkpatrick 1995). The initial study utilized a relatively heavy and complicated dart system. The loading of the vaccine into the dart required a special syringe, and then the dart had to be hand-pressurized with still another syringe. Barbs had to be removed before use because animals were not being immobilized. Finally, because the darts utilized needles with very small side-port openings, the viscous vaccine–adjuvant emulsion was injected slowly, and the dart often bounced out before the injection was complete. This difficulty led to the need to "lob" darts into the animals, so they would remain for a few seconds before falling out, which in turn restricted range to perhaps 25 m at best. By 1992, a new dart system was tested and found to be more versatile. The darts were smaller and lighter, could be loaded with a standard syringe, and did not require pressurization, relying instead on a powder charge that was detonated by impact. The smaller darts increased the range up to and beyond 50 m, and virtually all expended darts could be located and tested to determine if the vaccine had been injected.

The success on ASIS has led to application of PZP contraception for the wild horses on the Rachel Carson National Estuarine Reserve on Carrot Island, NC; the Shackleford Banks of Cape Lookout National Seashore, NC; Little Cumberland Island, GA; the Pryor Mountain National Wild Horse Refuge, MT; the Little Bookcliff National Wild Horse Range, CO; and the Return-to-Freedom Wild Horse Sanctuary in Lompoc, CA.

Although the PZP vaccine met the tests of the "ideal" characteristics as already outlined quite well, the necessity to administer two inoculations the first year was a definite hindrance. This limitation could be overcome in two ways. In the first case, tested successfully on ASIS in 1994–1995, a single inoculation was given to all untreated mares (n = 72) with no regard to contraceptive effects. A year later, when these same mares were given a single inoculation, contraception was better than 90 percent. Thus, the two-inoculation protocol can be avoided if the contraceptive effects during year 1 are ignored.

An alternative to this approach is to incorporate the PZP into some format that permits pulsed or delayed release. In an initial attempt, the PZP was incorporated into nontoxic biodegradable lactide-glycolide microspheres (prepared by D. Flanagan, University of Iowa), which on contact with water (that is, tissue fluids after injection) break down into lactic acid and carbon dioxide, releasing the PZP after some period of time. The actual release time can be altered by changing the ratio of lactide to glycolide. Initial tests with this delivery system took place with Nevada wild horses. One inoculation of a bolus of PZP plus Carbopol 934 (B. F. Goodrich) as an adjuvant plus PZP-containing microspheres, suspended in carboxymethylcellulose, resulted in a degree of fertility control that did not differ from the standard two-inoculation protocol (Turner et al. 1997, 2001). Although contraceptive results were promising, the microspheres settled out of the carrier medium, clogged syringes and needles, and made remote delivery with darts impossible.

The next step was to produce small cylindrical pellets of lactide glycolide that fit into the needle of a dart. One pellet contained 70 to 90 μg PZP and 150 to 175 μg QS-21 (Cambridge Biotech) as the adjuvant (a water-soluble saponin). A second pellet contained a larger dose of PZP, and a third contained 250 to 500 μg PZP. As the PZP–Freund's adjuvant bolus was pushed from the dart, the pellets would embed in the muscle tissue and release at some predetermined time. In January 2000, 96 free-ranging wild mares were treated with these pellets, and, in 2001, 6 percent of the treated mares (n = 32) produced foals whereas 59 percent of 33 untreated mares produced foals (Turner et al. 2002). After 2 years, the efficacy was 83 percent (J. W. Turner, personal communication).

A wild horse immunocontraceptive study in New Zealand conducted in the early 1990s illustrates the importance of both adjuvants and delivery systems (Cameron et al. 2001). The Kaimanawa wild horse herd inhabits a military reservation on central New Zealand's North Island. Twenty-six mares in nine different bands were treated with PZP, another 8 mares were given a placebo, and 63 mares were untreated to serve as controls. Seventeen of the treated mares received a second treatment about 2 months after the first. Each mare received 400 pg PZP incorporated into a biobullet, along with 20 μl synthetic trehalose dicorynomycolate (TCDM; RIBI ImmunoChem Research, 25 mg/ml) and squaline oil. The biobullet (BallistiVet) is a solid biodegradable projectile that is nontoxic and erodes on contact with tissue fluids.

Only a single vaccinated mare did not produce a foal a year later, and the causes for the failure were multiple. First, the choice of adjuvant probably resulted in poor immune responses. In an earlier controlled study with captive Kaimanawa mares (Stafford et al. 2001), immunization with PZP plus TCDM resulted in highly variable antibody titers in mares regardless of whether they received one, two, or three inoculations. Although the titers rose in treated mares, compared to control mares, they did not increase in steps after each new vaccination, as is commonly seen with better adjuvants, such as Freund's (Liu et al. 1989). Second, the delivery of the biobullets in the field, as opposed to corrals in the controlled study, was difficult because of limited range (20 m) and inability to determine accurately whether the animal was hit at all. Only if a trickle of blood at the target site was witnessed by the shooter was there any indication that the bullet had in fact hit the animal. Finally, the 400-pg PZP dose was considerably lower than the 65-μg dose used in the American studies. Collectively, the poor choice of adjuvant, the limitations posed by the biobullet compared to injecting darts, and the low PZP dose rendered this study unsuccessful, which highlights the importance of each parameter in conducting successful wild horse immunocontraceptive studies.

CERVIDS

The successful application of PZP to wild horses led directly to its use in zoo animals, and soon it was evident that this same vaccine worked in a variety of cervids (Kirkpatrick et al. 1996). In turn, this spawned interest in applying the vaccine to white-tailed deer inhabiting urban and suburban areas of the United States, where traditional lethal methods are no longer legal, wise, safe, or publicly acceptable. After several studies with captive deer (Turner et al. 1992, 1996b) that demonstrated the effectiveness and reversibility of PZP in white-tailed deer, a study was mounted with semifree-ranging deer at the Smithsonian Institute's Conservation and Research Center in Front Royal, VA. This study showed that two 65-μg PZP inoculations are necessary for contraception the initial year, that annual 65-μg PZP boosters will maintain the contraception, and that contraception resulted in an extended breeding season for this species, increasing it by as much as 2 months, a phenomenon that, as already discussed, does not occur in wild horses.

The study also demonstrated the same phenomenon that was seen on ASIS with wild horses; that is, that a single inoculation in the first year would not achieve appreciable levels of contraception, but a single inoculation a year later was successful in inhibiting fertility (McShea et al. 1997). There was no histological evidence of damage to the ovary after 2 years of treatment. These particular deer were extremely wary, so a variety of strategies had to be employed to inoculate all the animals in the study. However, despite the difficulties, it was demonstrated that wild deer could be treated successfully.

Three additional deer projects were mounted soon afterward. The first, which remains unpublished, involves deer within the Metro Parks system of Columbus, OH. Following a cull, PZP contraception has been applied every year since 1995, with favorable results in terms of preventing a rapidly growing population (J. W. Turner, personal communication). The most ambitious deer project was established on Fire Island National Seashore (FINS), NY, in 1993. The island, 55 km long, is a mosaic of established communities interspersed by national seashore, under the authority of the National Park Service. Deer populations within the communities were dense, and the problems were exacerbated by local residents feeding the deer large quantities of high-quality grains.

Because the deer were habituated to humans, several permanent residents of the island could identify deer using the same methods applied to wild horses on ASIS. From 1993 through 1997, 74 to 164 individually identified does were treated with 65 μg PZP plus FCA followed by booster inoculations of 65 μg PZP plus FCA, using blowguns. Between 1993 and 1997, fawning rates among individually identified animals declined by 78.9 percent from pretreatment levels. Population density in the most heavily treated area increased by 11 percent per year from 1995 through March 1998 and then declined by 23 percent per year through 2000 for a net reduction of almost 50 percent (Naugle et al. 2002). This study was the first demonstration that unrestrained deer populations could actually be reduced using contraception.

A third major deer project was initiated in 1993 on the grounds of the National Institute of Standards and Technology (NIST) in Gaithersburg, MD. This high-security facility, belonging to the US Department of Commerce, was inhabited by 185 deer in 1995 to a high of 300 deer in 1997. The deer were either immobilized, hand captured (in the case of fawns), or box-trapped and ear-tagged. Beginning in 1996, an initial 65 μg PZP plus FCA inoculation was given by hand injection, and boosters (same dose with FIA) were given by dart. During the course of the 6 years, a variety of other adjuvants was tested, including Carbapol, ISA-50 (Seppic), and RIBI adjuvant system. Carbapol is a long-chain carbohydrate; ISA-50 consists of 85 percent mineral oil and 15 percent mannide oleate; and the RIBI adjuvant system consists of monophosphoyl lipid A, synthetic trehalose dicorynomycolate, and bacterial cell-wall skeletons. Treated females produced a mean of 0.59 fawns per female in 1997, and then the mean declined to 0.17 to 0.26 fawns per female from 1998 through 2000. As the proportion of does receiving treatment increased to 80 percent in 1998, population fertility declined and the population declined from a high of 300 in 1997 to 218 by May 2001 (Rutberg et al. 2004).

In a slightly different approach, DeNicola et al. (1996) administered human chorionic gonadotropin (hCG: Sigma BioChemical) conjugated to ovalbumin/saponin to two herds of free-ranging white-tailed deer. The formulation, incorporated into biobullets (Antech), was delivered in October and November. All control and hCG-treated deer fawned the following spring. Twenty-one of 58 does were given booster inoculations of porcine luteinizing hormone (pLH: Sioux Biochemical) in the second year, in September and October, and efficacy was determined the subsequent spring. The boosted does actually had a higher fawning rate than the control does, suggesting that pLH enhanced reproduction.

In an attempt to induce abortion in two herds of semifree-ranging white-tailed does, DeNicola et al. (1997) administered 25 mg prostaglandin $F_{2\alpha}$ ($PGF_{2\alpha}$: Parke-Davis) in biobullets to 18 does in February (which is approximately 94 to 95 days into the approximate 200-day gestation). None of the 18 does produced fawns following treatment. In a repeat of that experiment, only 38 percent of another 8 does captured and treated in the same manner produced fawns, which was a significant reduction compared to control groups. These experiments were important in that they confirmed that $PGF_{2\alpha}$ was an effective abortifacient if given before day 150 of gestation, during which deer rely upon ovarian progesterone to maintain pregnancy. Earlier experiments with hand-injected captive deer using an aqueous solution of $PGF_{2\alpha}$ (Becker and Katz 1994) were unsuccessful. The interpretation of the success in the study by DeNicola et al. (1997) was that a slower release resulted from the biobullet delivery, exposing the corpus luteum to the $PGF_{2\alpha}$ for a longer period of time. The difference between the 100 percent efficacy among the 18 does treated remotely and the 62 percent efficacy among the captured animals was attributed to capture-related stress. Plotka et al. (1983) hypothesized that increased stress generates sufficient adrenal progesterone to sustain pregnancies.

Despite some very successful applications of contraception to wild deer populations, resistance from state fish and game agencies and a requirement by the Food and Drug Administration of treating only ear-tagged animals prevented a wider application of the technology to urban deer.

In 1995, resource managers at Point Reyes National Seashore, CA, began considering the use of contraception in its Tule wapiti herd (*Cervus elaphus nannodes*). These wapiti were the descendants of 10 animals translocated there in 1978, and by 1998 the herd had reached 549. As with most national parks hunting was not an option, and public opinion was not on the side of culling by the National Park Service (NPS). Beginning in 1997, 29 adult cows were initially treated with 100 μg PZP plus FCA by hand injection. Booster inoculations of 100 μg PZP plus

FIA were delivered remotely, by dart, 3 to 6 weeks later. In 1998 another 29 adult cows were treated in the same manner, and 15 of the 1997 cohort were treated with boosters. In 1999 the remaining 50 wapiti cows that had been treated in 1997 and 1998 were given boosters as described. The contraceptive efficacy for the 1997 and 1998 treated cohorts was 97 and 84 percent, respectively. Of the 14 cows that were not given booster inoculations in 1998, 36 percent had calves within 1 year of treatment. Untreated control calving rates during this time were 77 percent (Shideler et al. 2002).

In the only other contraceptive attempt with free-ranging wapiti, Heilmann et al. (1998) examined behavior changes among 10 radio-collared cow wapiti in Idaho. Animals were treated with 65 μg PZP plus FCA for the first inoculation and 65 μg ZP plus FIA for the booster inoculation. As expected on the basis of the McShea et al. (1997) study with deer, the treated wapiti extended their breeding behaviors, described as sexual interactions, well beyond the normal early fall breeding season. Beyond that, there were no changes in social structure or activity patterns.

FELIDS

Contraception has been attempted in a variety of wild felids. The justification for slowing population growth in animals such as the African lion (*Panthera leo*), cheetahs (*Acinonyx jubatus*), and leopards (*Panthera pardus*) is built on (1) the economics of smaller to medium-sized game parks in Africa and (2) changing agricultural practices, which in turn promote large increases in lion populations. As income from agriculture continues to fall, many ranches in South Africa have turned to ecotourism. A portion of the agricultural land is turned over to wild game populations, and an infrastructure is developed to house and transport tourists who are interested in viewing wildlife. A single pride of lions, or perhaps two prides, is tolerable with regard to the numerous ungulate populations inhabiting these small parks, and even desirable so far as tourists are concerned. However, as the lion populations grow, ungulate viewing opportunities decrease, and more animals must be translocated to the park, at great expense, to maintain adequate viewing opportunities.

In Namibia, changing agricultural practices have resulted in unprecedented increases in lion populations. Before the influence of modern man on agricultural practices, ungulate populations migrated large distances in response to drought and rainy seasons. Lions had to follow these migrations, resulting in high cub mortality. With the advent of fencing, artificial water sources, and the disruption of migrations, lion cub survival caused populations to rise to unprecedented and unacceptable levels (Berry 1996).

The first published attempts at applying fertility control to free-ranging lions occurred in Namibia's Etosha National Park in 1981. Based on previous successful experiments with captive lions (Seal et al. 1977), 10 lionesses were implanted with melengestrol acetate (MGA) Silastic implants, marked, and released. Over a 3-year period none of the implanted lionesses produced cubs. On removal of the implants, all returned to normal fertility. None of the implanted lionesses displayed any estrous behaviors while harboring the implants, implying ovulation failure (Orford 1978, 1996; Orford and Perrin 1988). Contraception was discontinued after 3 years because of a decline in lion populations for unrelated reasons. Dosages for the implants were not given in any of the publications, but previous work with zoo lions suggests that the implants contained approximately 500 mg MGA (Seal et al. 1977).

There are no other published reports of the use of MGA implants in free-ranging lions or other large cats, perhaps because of the reports that began emerging at about the same time of pathogenic effects of this steroid on the reproductive system of the recipient animals (Kollias et al. 1984; Linnehan and Edwards 1991; Vollset and Jakobsen 1986).

Orally delivered steroids have been used successfully to inhibit reproduction in feral domestic cats (*Felis catus*). In a field study in Scotland, an initial dose of 5.0 mg megestrol acetate was delivered in meat baits to 15 cats, followed by 2.5 mg per week thereafter. Only 4 of the 15 cats had litters (Remfry 1978). A repeat of this experiment in England was also successful (McDonald 1980).

More recently, interest has turned to GnRH agonists. In an initial experiment with deslorelin, 6-mg long-acting implants were placed in 4 female and 4 male cheetahs and 1 female leopard. Implants of 12 mg were placed in 2 lionesses living under semifree-ranging conditions (Bertschinger et al. 2001). None of the females became pregnant; although the 2 lionesses and the 4 cheetahs showed clinical signs of estrus, they did not permit mating. Within 12 to 18 months clinical signs of full estrus reappeared in 1 of the lionesses, suggesting reversibility of contraceptive action. The male cheetahs showed no evidence of sperm by day 82 after treatment, and 2 nontreated females living with these males did not become pregnant. The study was expanded to 31 cheetahs, 10 lionesses, and 4 leopards, which were either semifree ranging or completely free ranging, in Mabula Nature Reserve, RSA. The results were essentially the same (Bertschinger et al. 2002), and no behavioral side effects were noted. Because the implant matrix containing deslorelin degrades over time without the need for removal (Trigg et al. 2001), this approach may be ideal for free-ranging carnivores that can be captured only once for insertion.

ELEPHANTIDS

During the 1980s, a dramatic decrease in wild African elephant (*Loxodonta africana*) populations occurred, from an estimated 1.3 million in 1979 to about 600,000 in 1989, as a result of poaching for ivory and droughts (Poole 1994). Increased protection by organizations such as the Kenya Wildlife Service and the South African National Parks Board reversed these trends, and African elephant populations have been growing at significant rates since 1990. In Kenya growth rates have been estimated at 4 percent annually (Moss 1994), and in the Kruger National Park in South Africa growth rates have exceeded 5 percent (Joubert 1986). The years of heavy poaching have forced many elephant populations to seek refuge in smaller, compressed habitats, and at the same time much of their former habitat has been converted to agricultural use, which is largely incompatible with elephant populations. Now, despite elephant numbers being smaller than in the 1970s, population control methods, and in particular culling, have had to be implemented throughout many Africa nations.

In 1992, a pilot project was initiated in Kenya to test the effectiveness of the PZP contraceptive vaccine in free-roaming elephants. A small population was inoculated but the project was never completed because of lack of funding by the Kenya Wildlife Service. However, results from analysis of several blood samples indicated that the cows were making antibodies against the vaccine (B. Dunbar, personal communication).

Beginning in 1996, a large-scale project to test the contraceptive efficacy of the PZP vaccine was initiated in the Kruger National Park in South Africa. Ovaries recovered from culled elephants were incubated with rabbit anti-PZP-labeled antibodies, and binding indicated that there was significant cross-reactivity between the antibodies and the elephant zona pellucida (Fayrer-Hosken et al. 1999). In the initial experiment, 21 cow elephants were immobilized, radio-collared, and treated with 600 μg PZP in synthetic trehalose dicorynomycolate (S-TDCM; RIBI Immunochem Research). Inoculations were given on days 0, 30, and 270. The two booster inoculations were given remotely by dart from a helicopter. A year later 18 of 20 control cows were pregnant, whereas 9 of 19 treated cows were pregnant, representing a significant reduction in pregnancy rate. A second group of 10 cows were captured and treated at days 0, 14, and 42, and a year later only 2 were pregnant (Fayrer-Hosken et al. 2000).

To test both the extension of contraceptive effects through booster inoculations and the reversibility of contraceptive action, seven of the cows originally treated in 1996 were recaptured. Four were given 400 µg PZP plus S-TDCM booster inoculations and three were left untreated. A year later none of the four treated cows was pregnant, and ultrasound examinations indicated that all three cows that had not received booster inoculations were pregnant. Ultrasound images of ovaries and uteri of nonpregnant animals appeared normal, and the former showed evidence of healthy follicles and corpora lutea, indicating normal cycling activity (Fayrer-Hosken et al. 2000).

Based on the success of these trials with free-ranging elephants, a new project was initiated at Makalali Private Game Reserve, South Africa. In this project, all PZP vaccine was delivered remotely by dart, from the ground, and social behaviors and time budgets were studied (Delsink et al. 2002). Eighteen cows were treated with either 400 or 600 µg PZP plus Freund's modified adjuvant. Minor changes in family group movements and time budget utilization occurred immediately following treatment, but these returned to pretreatment values within 2 weeks after treatment. Four of the treated cows were pregnant at the time of treatment, and all produced healthy calves, indicating that PZP treatment did not interfere with the health of the calves in preexisting pregnancies. Preliminary results gathered in 2003 suggest that contraception is effective (A. Delsink, personal communication).

MACROPODS

Two different marsupials have presented Australians with overpopulation dilemmas. The eastern gray kangaroo (*Macropus giganteus*) is the counterpart to the urban white-tailed deer in North America in that it inhabits densely human-populated areas and industrial sites and represents a real or perceived nuisance, safety hazard, agricultural damage, and sometimes a threat to biodiversity. The koala (*Phascolarctos cinereus*) is found only in small numbers in some of its historic range, but on island and fragmented forest habitats it often exceeds habitat carrying capacity, thus requiring some form of population control. Additionally, brush-tailed possums (*Trichosurus vulpecula*) translocated to New Zealand represent a marsupial pest in that country.

Because the neuroendocrine control of folliculogenesis and ovulation is similar in marsupials and eutherians, the possibility of utilizing synthetic progestins to inhibit marsupial fertility was investigated. The tammar wallaby (*Macropus eugenii*) was selected as a model, and the synthetic progestin levonorgestrel (LNG) was administered in the form of Norplant implants (Leiras Pharmaceutical) (Nave et al. 2000). Each implant contained 70 mg LNG and was assumed to release the drug at the same rate as in women, 30 μg/day initially and then 15 mg/day after 12 months (Sivin 1988). Among control animals, 86 percent gave birth to new pouch young, 88 percent mated at the end of the first estrous cycle, and thereafter 94 percent produced young. Normal reproductive activity, defined as reproductive behaviors, breeding, and production of offspring, continued through 48 months. Among 24 implanted wallabies, only 42 percent gave birth at the end of the first cycle and none bred or produced young after that.

The same type of implant was next tested in a wild free-ranging population of eastern gray kangaroos. Seventeen kangaroos were captured and anesthetized, and each female received two implants subdermally. Eight control females were given empty rods, and all animals were released. A year later 7 of the control females (88 percent) produced young, and 3 of the treated females (19 percent) produced young (all 3 had pouch young at the time they were implanted, indicating that they were already in a diapause pregnancy). At the end of the second year all 8 control females and none of the treated females produced young (Nave et al. 2001, 2002; Poiani et al. 2002). There was no difference between the control and treated groups with regard to time spent feeding and grooming, although more males associated with control females.

PINNIPEDS

Porcine zona pellucida incorporated into multilaminar liposomes was administered to grey seals (*Halichoerus grypus*) on Sable Island, Nova Scotia. Sixty-four 14-year-old females, seventeen 20-year-old females, and twenty 21-year-old females were inoculated a single time with 100 μg PZP in 0.5-ml liposomes emulsified with 0.5 ml FCA. Over a 5-year period the birth rate, based on the return or failure to return to breeding grounds, dropped significantly among treated animals. Treated animals that did return were caught and bled, and antibody titers against the PZP were measured. The determined threshold for a contraceptive titer was 5 percent of reference serum standards, which is remarkably low compared to titers in other species (Brown et al. 1996, 1997a, 1997b).

RODENTS

The history of rodent contraception is another example of large numbers of studies occurring with captive wild or laboratory rodents under carefully controlled conditions with few actual published reports of application, or even study, with free-ranging populations. Some of the earliest attempts at wildlife contraception occurred with captive rodents, such as mice (*Mus* sp. and *Peromyscus* sp.) and voles (*Microtus* sp.), using a variety of steroid compounds (reviewed by Kirkpatrick and Turner 1984), but problems associated with bait acceptance, similar to those seen with deer, coyotes, and foxes, prevented field application. Several more recent studies have been conducted to assess the role of fertility control in rodents, but the methods used involved only castration or tubal ligations, which by definition are not contraception.

The only true contraceptive study with wild rodents occurred with the black-tailed prairie dog (*Cynomys ludovicanus*). Garrott and Franklin (1983) fed prairie dogs inhabiting Wind Cave National Park diethylstilbestrol (DES) incorporated into an oat bait. The DES was mixed with hulled oats at a concentration of 0.11 percent, and 25 g was placed at each active burrow twice per week, in March, just preceding the breeding season. After 1 year, 5 of 13 treated females (38 percent) produced litters, and after 2 years, 17 of 21 treated females (81 percent) produced litters. By year 3, 8 of 13 (62 percent) control females produced litters whereas no litters were produced by 21 treated females. In year 4, control reproductive success rose to 89 percent, and treated animals produced no litters at all. No later reports of this approach have appeared, probably because of the environmental issues surrounding the incorporation of a powerful antifertility steroid in prey species utilized by a variety of raptors and predatory mammals.

MUSTELIDS

Only two published reports exist for studies of contraception in wild mustelids. In the first study, DES-loaded egg baits were put out in striped skunk (*Mephitis mephitis*) habitat, but poor bait acceptance and erratic bait consumption led to unsuccessful results (Storm and Sanderson 1969). In a second study with free-ranging striped skunks, Norplant implants were placed in radio-collared females. One implant was placed in each skunk, and a year later none of the four recovered treated animals was pregnant or lactating (Bickle et al. 1991). Although this latter study was short term, it corroborates the effectiveness of Norplant implants in kangaroos (Nave et al. 2002) and the active ingredient levonorgestrol (LNG) in domestic cats (Looper at al. 2001).

This approach appears attractive for small mammals inhabiting urban areas, such as raccoons and skunks, where capture is relatively easy and where the chance for passage of the LNG to predator species is minimized. However, the possibility of deleterious effects of synthetic progestins in carnivores was not resolved by this study, and long-term investigations are needed. It has been estimated that skunks inhabiting urban areas do not normally live for more than 2 to 3 years; thus, any deleterious effects of progestins in this species in this setting may not be serious. The cost of the implants, however, probably precludes their widespread use in the near future.

PROCYONIDS

Only a single study exists for attempts at contraception in wild raccoons, that of Kramer (1996) using levonorgestrel (LNG) implants in free-ranging raccoons (*Procyon lotor*) on Canaveral National Seashore, FL. The implants were the same as those used in the skunk studies of Bickle et al. (1991) and the kangaroo studies of Nave et al. (2002), and each contained approximately 70 mg LNG with a release rate of 30 μg/day. The attempt failed because the application of this progestin, and probably others as well, can block parturition in the pregnant female, potentially causing LNG to be lethal if given to pregnant animals (see discussion in Chapter 7, Contraception in Carnivores). A previously unpublished study, conducted with LNG implants in captive raccoons, in tandem with the skunk study of Bickle et al. (1991), showed that LNG will inhibit fertility in nonpregnant raccoons (at a rate of two implants per animal). That study was conducted in Iowa, where raccoons are exclusively seasonal with regard to reproduction and where all treated animals were nonpregnant, whereas the raccoons on Canaveral National Seashore were not seasonal and some were pregnant at all times of the year. Thus, this approach might only be useful in more northern climates, where animals could be treated during the late summer and fall, when they were not pregnant. The LNG implants did not alter home range size, compared to controls. Although this experiment was not successful, the possibility of using LNG implants in urban raccoons in northern latitudes has some potential.

SUMMARY

The recent application of contraceptives to free-ranging wildlife management suggests fertility control is a reasonable approach to population management in some species. There is an impressive array of research in progress aimed at wildlife contraception, but most of this work is still at a stage where its use is confined to captive populations. As this research moves forward, the investigators must pay more attention to the problems and logistics of actual application in the wild or a great deal of effort and money will be lost. It is obvious that we need more and better contraceptive compounds and vaccines, but the greatest needs are for delivery systems and qualified and trained personnel to carry out the work in the field.

Delivery can be oral, by implant or hand injection after capture, or by dart in unrestrained populations, but the contraceptive agent must be compatible with the delivery system, and those problems must be addressed during development of the compounds and agents. Successful delivery of contraceptives to wild populations requires extraordinary understanding of the target species natural history and behavior as well as effective and safe dart delivery technology. Finally, not just anyone can move from the laboratory to the field and be successful. The qualities inherent in successful field operatives are often less tangible than those for a fine molecular biologist or immunologist.

The development of newer and better contraceptives must also be carried out in concert with regulatory agencies, such as the Food and Drug Administration or the US Department of Agriculture (USDA), or the results may be effective contraceptive technology that will not be permitted in the field (for discussion, see Chapter 2, Regulatory Issues). For example, it is unlikely that any orally delivered wildlife contraceptive will be licensed by a regulatory agency unless it is species specific, at least in the United States, yet research groups often pursue this approach without even a preliminary meeting with the regulators.

The economics of wildlife contraception pose special problems and hurdles to advancement. Most major pharmaceutical companies, or even small proprietary drug companies, must think in terms of sales of millions of doses per year to make acceptable profits. At the same time, the numbers inherent in wildlife contraception, far fewer than millions of doses per year, preclude the specific development of wildlife contraceptives for commercial reasons. No company will recover its investment if the only use for the contraceptive is in wildlife; thus, research advances will be slow unless there is an application for companion animals or domestic livestock as well. The cost of bringing a new drug or vaccine to commercial development is well beyond the expected returns from a successful product. Thus, wildlife contraceptive development will continue to creep along with nonprofit groups, or perhaps a bit faster in government-funded agencies such as the USDA or Commonwealth Scientific and Industrial Research Organisation (CSIRO) in Australia.

Finally, the entire subdiscipline of wildlife contraception is laced with political, social, cultural, and economic issues that can stop even the best technology dead in its tracks. There are ethical and moral issues surrounding this field, and those who nobly pursue more humane approaches to wildlife management may run afoul of one or more opposition groups, including animal rights groups, hunters, or those who cannot accept change. Ultimately, the scientists who pursue this goal, regardless of their altruism or lack thereof, must listen carefully to the true owners of all this wildlife—the public—if they are to be successful.

REFERENCES

Allen, S. H. 1982. Bait consumption and diethylstilbestrol influence on North Dakota red fox reproductive performance. *Wildl Soc Bull* 10:370–374.

Asa, C. S. 1999. Male reproductive success in free-ranging feral horses. *Behav Ecol Sociobiol* 47:89–93.

Balser, D. C. 1964. Management of predator populations with antifertility agents. *J Wildl Manag* 28:352–358.

Becker, S. E., and L. S. Katz. 1994. Effects of exogenous prostaglandin $F_{2\alpha}$ ($PGF_{2\alpha}$) on pregnancy status in white-tailed deer. *Zoo Biol* 13:315–323.

Bell, R. L., and T. J. Peterle. 1975. Hormone implants control reproduction in white-tailed deer. *Wildl Soc Bull* 3:152–156.

Berry, H. H. 1996. Ecological background and management application of contraception in free-living African lions. In *Contraception in Wildlife*, ed. P. N. Cohn, E. D. Plotka, and U. S. Seal, 321–338. Lewiston, NY: Mellon.

Bertschinger, H. J., C. S. Asa, P. P. Calle, J. A. Long, K. Bauman, K. Dematteo, W. Jöchle, T. E. Trigg, and A. Human. 2001. Control of reproduction and sex related behaviour in exotic wild carnivores with the GnRH analogue deslorelin; preliminary observations. *J Reprod Fertil Suppl* 57:275–283.

Bertschinger, H. J., T. E. Trigg, W. Jöchle, and A. Human. 2002. Induction of contraception in some African wild carnivores by down-regulation of LH and FSH secretion using the GnRH analogue deslorelin. *Reproduction Suppl* 60:41–52.

Bickle, C. A., J. F. Kirkpatrick, and J. W. Turner. 1991. Contraception in striped skunks with Norplant implants. *Wildl Soc Bull* 19:334–338.

Brown, R. G., W. C. Kimmins, M. Mezei, J. L. Parsons, B. Pohajdak, and W. D. Bowen. 1996. Birth control for grey seals. *Nature* (Lond) 379:30–31.

Brown, R. G., W. D. Bowen, J. D. Eddington, W. C. Kimmins, M. J. Medzei, J. L. Parsons, and B. Pohajdak. 1997a. Evidence for a long-lasting single administration contraceptive vaccine in wild grey seals. *J Reprod Immunol* 35:43–51.

Brown, R. G., W. D. Bowen, J. D. Eddington, W. C. Kimmins, M. J. Medzei, J. L. Parsons, and B. Pohajdak. 1997b. Temporal trends in antibody production in captive grey, harp and hooded seals to a single administration immunocontraceptive vaccine. *J Reprod Immunol* 35:53–64.

Brushman, H. H., S. B. Linhart, D. S. Balser, and L. W. Sparks. 1967. A technique for producing antifertility tallow baits for predatory mammals. *J Wildl Manag* 32:183–184.

Cameron, E. Z., W. L. Linklater, E. O. Minot, and K. J. Stafford. 2001. *Population Dynamics 1994–1998, and Management of the Kaimanawa Wild Horses*. Science Conservation Bulletin 171. Wellington, NZ: Department of Conservation.

Cheatum, E. L., and W. Hansel. 1967. Rabies control by inhibition of fox reproduction. Master's thesis. Ithaca: Cornell University.

Delsink, A., J. J. Van Altena, J. F. Kirkpatrick, D. Grobler, and R. A. Fayrer-Hosken. 2002. Field application of immunocontraception in African elephants (*Loxodonta africana*). *Reproduction Suppl* 60:117–124.

DeNicola, A. J., R. W. Swihart, and D. J. Kesler. 1996. The effect of remotely-delivered gonadotropin formulations on reproductive function of white-tailed deer. *Drug Dev Ind Pharm* 22:847–850.

DeNicola, A. J., D. J. Kesler, and R. W. Swihart. 1997. Remotely delivered prostaglandin $F_{2\alpha}$ implants terminate pregnancy in white-tailed deer. *Wildl Soc Bull* 25: 527–531.

Eagle, T. C., E. D. Plotka, R. A. Garrott, D. B. Siniff, and J. R. Tester. 1992. Efficacy of chemical contraception in feral mares. *Wildl Soc Bull* 20:211–216.

Fayrer-Hosken, R. A., H. J. Bertschinger, J. F. Kirkpatrick, D. Grobler, N. Lamberski, G. Honneyman, and T. Ulrich. 1999. Contraceptive potential of the porcine zona pellucida vaccine in the African elephant (*Loxodonta africana*). *Theriogenology* 52:835–846.

Fayrer-Hosken, R. A., D. Grobler, J. J. Van Altena, H. J. Bertschinger, and J. F. Kirkpatrick. 2000. Immunocontraception of African elephants. *Nature* (Lond) 407:149.

Garrott, M. G., and W. L. Franklin. 1983. Diethylstilbestrol as a temporary chemosterilant to control black-tailed prairie dog populations. *J Range Manag* 36:753–756.

Gates, N. L., C. S. Card, V. Eroschenko, and C. V. Hulet. 1976. Insensitivity of the coyote testis to orally administered cadmium. *Theriogenology* 5:281–286.

Goodloe, R. B., R. J. Warren, and D. C. Sharp. 1996. Immunization of feral and captive horses: a preliminary report. In *Contraception in Wildlife*, ed. P. N. Cohn, E. D. Plotka, and U. S. Seal, 229–242. Lewiston, NY: Mellon.

Greer, K. R., W. H. Hawkins, and J. E. Catlin. 1968. Experimental studies of controlled reproduction in elk (wapiti). *J Wildl Manag* 32:368–376.

Harder, J. D. 1971. The application of an antifertility agent in the control of a white-tailed deer population. Ph.D. dissertation. Columbus: The Ohio State University.

Harder, J. D., and T. J. Peterle. 1974. Effects of diethylstilbestrol on reproductive performance in white-tailed deer. *J Wildl Manag* 38:183–196.

Heilman, T. J., R. A. Garrott, L. L. Cadwell, and B. L. Brett. 1998. Behavioral response of free-ranging elk treated with an immunocontraceptive vaccine. *J Wildl Manag* 62:243–250.

Herzog, H. A., A. Rowan, and D. Kossow. 2001. Social attitudes and animals. In *The State of the Animals 2001*, ed. D. J. Salem and A. Rowan, 55–70. Washington, DC: Humane Society Press.

Joubert, S. C. J. 1986. Master plan for the management of the Kruger National Park. Unpublished internal memorandum. Skukuza, Republic of South Africa: National Parks Board.

Kirkpatrick, J. F. 1995. *Management of Wild Horses by Fertility Control: the Assateague Experience.* National Park Service Scientific Monograph. Denver, CO: National Park Service.

Kirkpatrick, J. F., and J. F. Turner. 1984. Chemical fertility control and wildlife management. *BioScience* 35:485–491.

Kirkpatrick, J. F., and J. W. Turner. 1991. Reversible contraception in nondomestic animals. *J Zoo Wildl Med* 22:392–408.

Kirkpatrick, J. F., and J. W. Turner. 1995. Urban deer fertility control: scientific, social and political issues. *Northeast Wildl* 52:103–116.

Kirkpatrick, J. F., and A. Turner. 2002. Reversibility and safety during pregnancy in wild mares treated with porcine zona pellucida. *Reproduction Suppl* 60:197–202.

Kirkpatrick, J. F., and A. Turner. 2003. Absence of effects from immunocontraception on seasonal birth patterns and foal survival among barrier island wild horses. *J Appl Anim Welf* 6:301–308.

Kirkpatrick, J. F., J. W. Turner, and A. Perkins. 1982. Reversible fertility control in feral horses. *J Equine Vet Sci* 2:114–118.

Kirkpatrick, J. F., I. K. M. Liu, and J. W. Turner. 1990. Remotely-delivered immunocontraception in feral horses. *Wildl Soc Bull* 18:326–330.

Kirkpatrick, J. F., I. K. M. Liu, J. W. Turner, and M. Bernoco. 1991. Antigen recognition in feral mares previously immunized with porcine zona pellucida. *J Reprod Fertil Suppl* 44:321–325.

Kirkpatrick, J. F., I. K. M. Liu, J. W. Turner, R. Naugle, and R. Keiper. 1992. Long-term effects of porcine zonae pellucidae immunocontraception on ovarian function in feral horses (*Equus caballus*). *J Reprod Fertil* 94:437–444.

Kirkpatrick, J. F., R. Naugle, I. K. M. Liu, and J. W. Turner. 1995. Effects of seven consecutive years of porcine zona pellucida immunocontraception on ovarian function in feral mares. *Biol Reprod Monogr Ser 1 Equine Reprod* VI:411–418.

Kirkpatrick, J. F., P. P. Calle, P. Kalk, I. K. M. Liu, and J. W. Turner. 1996. Immunocontraception of captive exotic species. II. Formosa sika deer (*Cervus nippon taiounus*), Axis deer (*Cervus axis*), Himalayan tahr (*Hemitragus jemlahicus*), Roosevelt elk (*Cervus elaphus roosevelti*), Reeve's muntjac (*Muntiacus reevesi*) and sambar deer (*Cervus unicolor*). *J Zoo Wildl Med* 27:482–495.

Kollias, G. V., M. B. Calderwood-Mays, and B. G. Short. 1984. Diabetes mellitus and abdominal adenocarcinoma in a jaguar receiving megestrol acetate. *J Am Vet Med Assoc* 185:1383–1386.

Kramer, M. T. 1996. Reproductive biology and preliminary investigations of levonorgestrel as a contraceptive for managing raccoons at Canaveral National Seashore, Florida. M. Sc. thesis. Athens: University of Georgia.

Levenson, T. 1984. Family planning for deer. *Environment* 1984:35–38.

Linnehan, R. M., and J. L. Edwards. 1991. Endometrial adenocarcinoma in a Bengal tiger (*Panthera tigris bengalensis*) implanted with melengestrol acetate. *J Zoo Wildl Med* 22:358–363.

Linhart, S. B. 1963. Control of sylvatic rabies by antifertility agents: principles and problems. Presented at the Northwestern Fish and Wildlife Conference, Portland, ME.

Linhart, S. B. 1964. Acceptance by wild foxes of certain baits for administering antifertility agents. *NY Fish Game J* 11:69–77.

Linhart, S. B., and R. K. Enders. 1964. Some effects of diethylstilbestrol in captive red foxes. *J Wildl Manag* 28:358–363.

Liu, I. K.M., M. Bernoco, and M. Feldman. 1989. Contraception in mares heteroimmunized with porcine zona pellucida. *J Reprod Fertil* 85:19–29.

Looper, S., G. Anderson, Y. Sun, A. Shukla, and B. L. Lasley. 2001. Efficacy of levonorgestrol when administered as an irradiated slow-release injectable matrix for feline contraception. *Zoo Biol* 20:407–421.

Matschke, G. H. 1977. Microencapsulated diethylstilbestrol as an oral contraceptive in white-tailed deer. *J Wildl Manag* 41:87–91.

McDonald, M. 1980. Population control of feral cats using megestrol acetate. *Vet Rec* 106:129.

McShea, W. J., S. L. Monfort, S. Hakim, J. F. Kirkpatrick, I. K. M. Liu, J. W. Turner, L. Chassy, and L. Munson. 1997. The effect of immunocontraception on the behavior and reproduction of white-tailed deer. *J Wildl Manag* 61:560–69.

Moss, C. J. 1994. Some reproductive parameters in a population of African elephants, *Loxodonta africana*. In *Proceedings, Second International Conference on Human and Animal Reproduction*, ed. C. S. Bambra, 284–292. Nairobi, Kenya: Institute of Primate Research, National Museum of Kenya.

Naugle, R. E., A. T. Rutberg, H. B. Underwood, J. W. Turner, and I. K. M. Liu. 2002. Immunocontraception of white-tailed deer on Fire Island National Seashore, New York. *Reproduction Suppl* 60:143–153.

Nave, C. D., G. Shaw, R. V. Short, and M. B. Renfree. 2000. Contraceptive effects of levonorgestrel implants in a marsupial. *Reprod Fertil Dev* 12:81–86.

Nave, C. D., G. Coulson, A. Poiani, G. Shaw, and M. B. Renfree. 2001. Fertility control in the eastern grey kangaroo using levonorgestrel implants. *J Wildl Manag* 66: 59–66.

Nave, C. D., G. Coulson, R. V. Short, A. Poiani, G. Shaw, and M. B. Renfree. 2002. Levonorgestrel implants provide long-term fertility control in the kangaroo and wallaby. *Reproduction Suppl* 60:71–80.

Oleyar, C. M., and B. S. McGinnes. 1974. Field evaluation of diethylstilbestrol for suppressing reproduction in foxes. *J Wildl Manag* 38:101–106.

Orford, H. J. L. 1978. Reproductive physiology and hormonal contraception in free-ranging lions (*Panthera leo*) at the Etosha National Park. M. Sc. thesis. Natal University, Republic of South Africa.

Orford, H J. L. 1996. Hormonal contraception in free-ranging lions (*Panthera leo* L.) at the Etosha National Park. In *Contraception in Wildlife*, ed. P. N. Cohn, E. D. Plotka, and U. S. Seal, 303–320. Lewiston, NY: Mellon.

Orford, H. J. L, and M. R. Perrin. 1988. Contraception, reproduction and demography of free-ranging Etosha lions (*Panthera leo*). *J Zool* (Lond) 216:717–733.

Plotka, E. D., and D. N. Vevea. 1990. Serum ethinylestradiol (EE2) concentrations in feral mares following hormonal contraception with homogeneous Silastic implants. *Biol Reprod* 42(Suppl 1):43.

Plotka, E. D., U. S. Seal, J. L. Verme, and J. J. Ozaga. 1983. The adrenal gland in white-tailed deer: a significant source of progesterone. *J Wildl Manag* 47:38–44.

Poiani, A., G. Coulson, D. Salamon, S. Holland, and C. Nave. 2002. Fertility control of eastern grey kanagaroos: do levonorgestrel implants affect behavior? *J Wildl Manag* 66:59–66.

Poole, J. H. 1994. Logistical and ethical considerations in the management of elephant populations through fertility control. In *Proceedings, 2nd International Conference on Human and Animal Reproduction*, ed. C. S. Bambra, 278–283. Nairobi, Kenya: Institute of Primate Research, National Museum of Kenya.

Remfry, J. 1978. Control of feral cat populations by long-term administration of megestrol acetate. *Vet Rec* 103:403–404.

Roughton, R. D. 1979. Effects of oral melengestrol acetate on reproduction in captive white-tailed deer. *J Wildl Manag* 43:428–436.

Rutberg, A. T., R. E. Naule, L. S. Thiele, and I. K. M. Liu. 2004. Effects of immunocontraception on a suburban population of white-tailed deer *Odocoileus virginianus*. *Biol Conserv* 116:243–250.

Seal, U. S. 1991. Fertility control as a tool for regulating captive and free-ranging wildlife populations. *J Zoo Wildl Med* 22:1–5.

Seal, U. S., R. Barton, L. Mather, K. Oberding, E. D. Plotka, and C. W. Gray. 1977. Hormonal contraception in female lions (*Panthera leo*). *J Zoo Anim Med* 7:1–17.

Shideler, S. E., M. A. Stoops, N. A. Gee, J. A. Howell, and B. L. Lasley. 2002. Use of porcine zona pellucida vaccine as a contraceptive agent in free-ranging tule elk (*Cervus elaphus nannodes*) at Tomales Point, Point Reyes National Seashore. *Reproduction Suppl* 60:169–176.

Sivin, I. 1988. International experience with Norplant and Norplant-2 contraceptives. *Stud Fam Plann* 19:81–94.

Stafford, K. J., E. O. Minot, W. L. Linklater, E. Z. Cameron, and S. E. Todd. 2001. *Use of an Immunocontraceptive Vaccine in Feral Kaimanawa Mares*. Conservation Advisory Science Notes 330. Wellington, NZ: Department of Conservation.

Storm, G. L., and G. C. Sanderson. 1969. Results of a field test to control striped skunks with diethylstilbestrol. *Trans Ill State Acad Sci* 62:193–197.

Thompson, L. H. 1976. Induced sterility for coyote control: effect of cadmium chloride on potential fertility of the male *Canis familiaris*. *Sci Biol J* March–April: 42–47.

Trigg, T. E., P. J. Wright, A. F. Armour, P. E. Williamson, A. Junaidi, G. B. Martin, A. G. Doyle, and J. Walsh. 2001. Use of a GnRH analogue implant to produce reversible, long-term suppression of reproductive function of male and female domestic dogs. *J Reprod Fertil Suppl* 57:255–261.

Turner, J. W., and J. F. Kirkpatrick. 1982. Androgens, behavior and fertility control in feral stallions. *J Reprod Fertil Suppl* 32:79–87.

Turner, J. W., and J. F. Kirkpatrick. 2002. Population effects, increased longevity and condition among immunocontracepted wild mares (*Equus caballus*). *Reproduction Suppl* 60:187–195.

Turner, J. W., I. K. M. Liu, and J. F. Kirkpatrick. 1992. Remotely-delivered immunocontraception in captive white-tailed deer. *J Wildl Manag* 56:154–157.

Turner, J. W., I. K. M. Liu, and J. F. Kirkpatrick. 1996a. Remotely-delivered immunocontraception in free-roaming feral burros. *J Reprod Fertil* 107:31–35.

Turner, J. W., J. F. Kirkpatrick, and I. K. M. Liu. 1996b. Effectiveness, reversibility and serum antibody titers associated with immunocontraception in captive white-tailed deer. *J Wildl Manag* 60:45–51.

Turner, J. W., I. K. M. Liu, A. T. Rutberg, and J. F. Kirkpatrick. 1997. Immunocontraception limits foal production in free-roaming feral horses in Nevada. *J Wildl Manag* 61:873–880.

Turner, J. W., I. K. M. Liu, D. R. Flanagan, A. T. Rutberg, and J. F. Kirkpatrick. 2001. Immunocontraception in feral horses: one inoculation provides one year of infertility. *J Wildl Manag* 65:235–241.

Turner, J. W., I. K. M. Liu, D. R. Flanagan, K. S. Bynum, and A. T. Rutberg. 2002. PZP immunocontraception of wild horses in Nevada: a 10-year data base. *Reproduction Suppl* 60:177–186.

Vollset, I., and G. Jakobsen. 1986. Feline endocrine alopecia-like disease probably induced by medroxyprogesterone acetate. *Feline Pract* 16:16–17.

CHERYL S. ASA AND INGRID J. PORTON

EPILOGUE

Future Directions in Wildlife Contraception

Contraceptive use began in zoos more than 25 years ago when Dr. Ulysses Seal introduced the melengestrol acetate (MGA) implant for female lions and tigers. Since then, it has been effective in preventing reproduction in numerous mammalian species. The conspicuous exception is horses, and possibly other equids. No progestin other than altrenogest appears to be able to affect the endocrine physiology of horses (Plotka et al. 1988; Webel and Squires 1982). Unplanned pregnancies have occurred in one family of New World monkeys (Cebidae) with MGA implants, but the problem appears to be one of dose, not method. Very high doses of MGA or of Depo-Provera (Pharmacia and Upjohn) seem to be required to achieve hormone suppression by negative feedback, probably because New World monkeys have such high levels of naturally occurring sex steroid hormones, as much as 10 times higher than their cousins, the Old World monkeys (Coe et al. 1992). Regardless of the reason, however, MGA has proven to be a remarkably successful contraceptive, especially considering the huge range of species represented (see Chapters 7 through 11 on contraceptive application).

However, even though MGA may be effective across almost all mammalian taxa, it is not uniformly safe (see Chapter 5, Adverse Effects of Contraceptives), and recently questions about its reversibility in some species have been raised (De Vleeschouwer et al. 2000; DeMatteo et al. 2002). The taxonomic group that appears to experience the most deleterious effects from MGA and other progestins are the carnivores (Asa and Porton 1991). Although most investigations and reports concern felids and canids, the cautious assumption is that carnivores in general may be susceptible to these side effects, which target primarily their reproductive tissues, such as uterine lining and mammary glands.

Particularly because of this pressing need for safer contraceptives for carnivores, but also because managers of captive and free-ranging wildlife would benefit from more choices, the American Zoo and Aquarium Association (AZA) Wildlife Contraception Center gives high priority to the search for new contraceptive approaches. In the past, identifying promising new contraceptives for wildlife depended almost entirely on advances in research and development of contraceptives destined for the human market, because the high cost of pharmaceutical research and development restricts such efforts to markets more lucrative than those for wildlife. Unfortunately, support for human contraceptive development in the United States has been weak for more than a decade, with the result that few new approaches are being proposed and made available for testing. However, more recently, the field of contraceptive research has benefited by the growing interest in nonlethal approaches to reducing or managing populations of free-ranging wildlife, as well as stray and feral domestic cats and dogs, and in providing alternatives to surgical spaying and neutering for pets to help reduce the number relinquished to shelters.

The stimulation of interest in and financing of contraceptive research for dogs and cats may benefit wild carnivores as well. The interest in limiting free-ranging populations with contraception has focused primarily on ungulates. Even though the need for alternative contraceptives for captive ungulates is not as urgent as that for carnivores, such research may produce valuable alternatives for them also.

It is interesting that although most contraceptive methods currently available are for females, most of those in development are directed to males. This approach is especially good news for managers of captive wildlife because there have been so few options for treating males. In managing reproduction in captive wildlife, it is often more practical to contracept males, as the majority of mammalian species are polygynous; that is, in the natural social system one male mates with several females. In such a social group, it is only necessary to treat that one male to prevent reproduction in the entire group. Of course, with free-ranging wildlife, this can work against the manager for more promiscuous species such as white-tailed deer: Just one male deer that remains fertile can inseminate numerous females.

The following review focuses on the most promising new methods, those that seem most suitable for wildlife application and that have advanced past the initial stages of investigation. Some are being developed primarily for domestic and feral dogs and cats, and some are intended for human application, but all could potentially be used in either captive or free-ranging wildlife.

FEMALE-DIRECTED METHODS

Hormonal Methods

STEROID HORMONES: PROGESTINS To address problems with delivery, yet another vehicle containing MGA is planned (a project of the AZA Wildlife Contraception Center). Currently incorporated in Silastic rod implants and in commercial herbivore diets, MGA will also be available in a syrup (Zoo-Pharm) that can be added to an animal's favorite treat to encourage ingestion.

A new synthetic progestin, Nestorone (The Population Council, New York), is being developed as an implant (Massai et al. 2001). Its advantage over other progestins is that it is not orally active and so would be an especially suitable contraceptive for lactating females. Even though it may be transferred to nursing infants in milk, the hormone is not absorbed from the digestive tract and thus cannot enter the infant's system. No deleterious effects of progestin contraception on infant growth and development have been found (WHO 1994a, 1994b), but preventing passage of the hormone to infants provides an extra measure of certainty. A progestin that is not orally active might also be useful as a contraceptive for free-ranging wildlife that are currently ineligible for such treatment because of the potential for entering the food supply for humans.

STEROID HORMONES: ESTROGEN–PROGESTIN COMBINATIONS In addition to the well-known birth control pills and the newer depot injections containing both an estrogen and progestin (see Chapter 3, Types of Contraception), a vaginal ring containing ethinyl estradiol and etonorgestrel (NuvaRing) was introduced in the United States in 2002 by Organon. Although it has not been tested in wildlife, it might be appropriate for some species. If it prevents estrus, so that copulation does not occur, the ring might stay in place in females with a vaginal vault approximately the size of the human vagina. It might not be suitable for most primates, however, because they all have much more shallow vaginas than do humans; in addition, their manual dexterity would facilitate their ability to reach and remove the ring.

Immunocontraception

ANTISPERM VACCINES The presence of antisperm antibodies in human males that affect fertility suggested the feasibility of an antisperm contraceptive vaccine (Gupta et al. 1975). In contraceptive development, it has been more effective to immunize females rather than males against sperm. Although several antigens are promising (Primakoff et al. 1988; Zhu and Naz 1997; Hardy et al. 1997; Lea et al. 1998), none has yet been approved for commercial use. An interesting solution to the considerable individual variability in time to reversal with vaccines involved neutralizing antisperm antibodies by intravaginal administration of the antigen (Naz and Chauhan 2002), because a major site of antibodies is in cervical mucus (Mazumdar and Levine 1998).

VACCINES AGAINST EGG-SPECIFIC PROTEINS Components of the egg other than the zona pellucida (ZP) glycoproteins targeted by ZP vaccines are the focus of study for potential contraceptive development (Coonrod 2002). DNA libraries of dog and cat ovarian tissue have been established and the process of isolating proteins has begun, but none has yet been tested.

PERMANENT METHODS FOR PREVENTING REPRODUCTION IN MALES

Chemical Castration

Early formulations for inducing testicular atrophy by direct injection into the testis were associated with considerable pain, even resulting in sloughing of the skin around the injection site. A more recent formulation tested in dogs, zinc gluconate followed with neutralization by arginine to control the extent of atrophy, is claimed to be free of painful side effects (Wang 2002). However, it seems to present little or no advantage over surgical castration for captive wildlife.

REVERSIBLE CONTRACEPTION FOR MALES

Reversible Vasectomy

Recent successful reversal of vasectomies in bush dogs (S. J. Silber, K. E. DeMatteo, R. E. Junge, C. Dutton, I. J. Porton, and C. S. Asa; unpublished observations) provides an encouraging model that may extend to other species. The likelihood of reversal was enhanced by performing the vasectomies with the "open-end" technique developed by Silber (1979), which facilitates reanastomosis. Leaving the end of the vas nearest the testis open prevents the damage that can result from back pressure produced as sperm are released from the testis into a blocked vas. There will be challenges in adapting the technique to other species, because vasa can differ substantially among species, but this initial success demonstrates the potential of the approach for achieving reversible contraception in males.

Hormonal Methods: Androgens

Development of a steroid-based contraceptive in males has concentrated primarily on testosterone and other androgens. Although sperm production is reduced with such treatment, azoospermia, the complete absence of sperm preferred to ensure infertility, has been possible in human males in only 60 to 90 percent of subjects, which has been deemed unacceptable. The accompanying supraphysiological levels of androgen have been associated with side effects such as changes in lipid profiles, weight gain, and acne. An additional obstacle has been the frequency of administration, as often as once weekly (WHO 1990, 1996). New androgen formulations (testosterone undecanoate, testosterone buciclate, and 7α-methyl-19-nortestosterone) now available are active for a month or more.

The major argument against using androgens is the potential for adverse health effects. Among those documented are deleterious effects on the liver, kidney, and the immune and cardiovascular systems (Kibble and Ross 1987; Middleman and DuRant 1996). Progestin alone has not been effective at suppressing spermatogenesis, even at doses several times higher than those sufficient to suppress ovulation in females. However, the combination of a synthetic progestin and androgen is more effective than either hormone alone at suppressing spermatogenesis; furthermore, the addition of progestin reduces the amount of androgen needed to suppress spermatogenesis and concomitantly reduces the androgen-associated side effects (Anderson et al. 2002; Meriggiola et al. 2002). However, progestin-related side effects such as decreased high-density lipoprotein (HDL) cholesterol and increases in body weight, hematocrit (Meriggiola et al. 2002), systolic and diastolic blood pressure, and liver enzymes prompted termination of the study (Anderson et al. 2002). The antiandrogen cyproterone acetate, not currently approved for use in the United States, when added to testosterone has resulted in the fewest side effects (Meriggiola et al. 1998).

An additional concern about androgen treatment is the possible stimulation of the prostate gland. Testosterone administered with an enzyme that prevents its conversion to dihydrotestosterone, the androgen that directly stimulates the prostate, has been one approach to protecting the prostate (McLachlan et al. 2000). Another new androgen, 17α-methyl-19-nortestosterone (MENT), which cannot be converted to dihydrotestosterone, is an alternative way to address that concern. Either alone or in combination with estradiol, MENT can suppress both testosterone and spermatogenesis (Ramachandra et al. 2002). However, none of these androgen contraceptives has been adequately evaluated for potential affects on aggression. Increased aggression has been documented not only in men taking synthetic androgens for muscle development but also in men treated for other clinical conditions such as aplastic anemia (Malone et al. 1995).

The most appropriate use of synthetic androgens may be in combination with a gonadotropin-releasing hormone (GnRH) agonist to maintain the testosterone-dependent secondary sex characteristics and male behavior. If testosterone concentrations were maintained at normal physiological levels, any deleterious effects of testosterone should not exceed those seen in untreated males. Even better, if a synthetic androgen that did not stimulate the prostate gland was used as the replacement, health risks should actually be lower than in untreated males. However, although testosterone suppression may be an issue in humans, it is less likely to be so for wildlife, except perhaps for species such as lions that might otherwise lose their manes.

Immunocontraceptives: Gonadotropin-Releasing Hormone Vaccines

The subject of research efforts for many years, GnRH vaccine development has progressed somewhat. Early preparations provoked only weak antibody responses, but newer versions appear more effective (Robbins 2002). In Australia, Improvac (CSL Animal Health) is marketed primarily for use in domestic boars to prevent the meat of adult males from being tainted with the odor stimulated by testosterone (Metz et al. 2002). Recent approval of a GnRH vaccine in the United States for the treatment of prostate cancer in dogs (Canine Gonadotropin Releasing Factor Immunotherapeutic: Bicor Animal Health) may also be applicable as a contraceptive in females and in other species.

Chemosterilants

BISDIAMINES A class of compounds that came close to being marketed as "the pill" for men has potential for contraception in other mammals. Bisdiamines reversibly inhibit spermatogenesis without affecting hormone levels (Beyler et al. 1961) but also inhibit the activity of alcohol dehydrogenase (Heller et al. 1961), the liver enzyme that detoxifies alcohol. The problem for commercial development for men is obvious, but for treatment of captive animals with no access to alcohol, this side effect presents no concern. The bisdiamine WIN 18,446 has undergone the most extensive testing and has been shown to be safe and effective in rhesus monkeys (*Macaca mulatta*), humans, several rodent species, dogs, and gray wolves (*Canis lupus*) (Drobeck and Coulston 1962; Asa et al. 1996). Because of the large dose necessary for efficacy and the associated costs, however, commercial development is not feasible.

INDENOPYRIDINE First tested as an antihistamine, the indenopyridine Sandoz 20-438 reversibly disrupts spermatogenesis in dogs, rats, and mice when administered orally without affecting testosterone levels (Hodel and Suter 1978; Matter et al. 1979). Twenty-eight days of treatment with another analogue were followed by no signs of pathology or mutagenicity, but mild lethargy (as with many antihistamines) associated with some weight loss occurred, although these effects subsided after the first 10 days of treatment, when tolerance developed (Fail et al. 2000). More testing is planned, but its efficacy in dogs suggests it might have application at least to canids. However, pilot tests with domestic cats resulted in diarrhea, sometimes tinged with blood (C. Asa, D. Kunze, and P. Fail; unpublished data), which may indicate it is not suitable for felids.

GOSSYPOL Gossypol, a polyphenol isolated from the cotton plant, can arrest spermatogenesis without affecting testosterone production following daily ingestion (Liu 1987). Early concerns about toxicity (hypokalemia, low blood potassium) and about irreversibility in about 10 percent of subjects made it unattractive for development (Weinbauer et al. 1983). However, recent investigations have challenged those results, at least in terms of hypokalemia, although reversibility remains a problem in up to 20 percent of subjects (Coutinho 2002). However, in free-ranging wildlife this level of permanent sterility might be acceptable, and in some captive males reversibility might not be critical.

INDAZOLE CARBOXYLIC ACIDS Two new analogues of lonidamine, an indazole carboxylic acid, induce premature release of immature sperm cells, rendering them incompetent to achieve fertilization (Cheng et al. 2002; Gatto et al. 2002). First developed as a class of anticancer drugs, these agents have recently been shown to be effective and reversible contraceptives that do not affect hormone secretion. However, toxicology studies have not yet been completed.

Mechanical Devices: Vas Plugs

A variety of devices, including injectable silicone, cylindrical plugs, spherical beads, threads of silicone or suture material, and series of polypropylene beads, have been tested (Lee 1969; Farcon et al. 1975; Mohr and Johnson 1978, Zaneveld et al. 1988; C. Asa and L. Zaneveld, unpublished data). However, none that were adequately evaluated sustained effective sperm obstruction without also causing scarring or perforation. This problem may be solved by injectable materials that solidify in place (Lohiya et al. 1998; Manivannan et al. 1999). One of these, the RISUG (Reversible Inhibition of Sperm Under Guidance), is in Phase III clinical trials (Chaki et al. 2003). It consists of a copolymer (styrene maleic anhydride) dissolved in dimethyl sulfoxide (DMSO) that does not have tissue sclerosing or adherent properties but is hydrophilic and porous, which results in swelling that lodges it in place. It inhibits fertility not only through structural blockage of the vas but also because the charged polymer damages the sperm membrane so that acrosin and hyaluronidase leak out, rendering the cell incapable of fertilization.

OBSTACLES TO DEVELOPMENT OF NEW CONTRACEPTIVES

Although the zoo and wildlife communities can benefit from these efforts, the major applications for dogs and cats require permanent inhibition of reproduction. Permanent sterilization may be appropriate for some free-ranging wildlife populations, but for the most part zoos need reversible methods for use in managed breeding programs. A further handicap to commercial development of animal contraceptives is the limited market, both in numbers and in potential profits, compared to the market for human contraceptives.

Pharmaceutical companies have little incentive to develop products when they are unlikely to recover their research and development costs from sales. The market for zoo and wildlife contraceptives is minuscule compared to that for humans or even perhaps for domestic animals. It is estimated that to get a new drug approved can cost a pharmaceutical company as much as $20 million and 8 to 10 years of research (Bren 2002), and other estimates are even higher. Then, FDA requirements for Good Manufacturing Practice (GMP) during production add still more to the per unit cost. Even if a contraceptive gained approval, the manufacturer would face the further challenge of producing it at a price affordable to zoo and wildlife managers. The result, under current FDA policy, is that even the most promising new contraceptives may never become available for wildlife application because of the prohibitive expense.

Because the expense entailed in developing and producing drugs for veterinary use restricts the products available to other groups as well, some groups have formed alliances to facilitate communication about promising new treatments, to jointly solicit funding to support research, and in some cases to lobby the FDA and Congress about the burden current policies place on efforts to improve animal care and well-being. Examples include the Alliance for Contraception in Cats and Dogs, which hosted a symposium in 2002 on nonsurgical methods for pet population control, bringing researchers together with foundations that are sympathetic to the pet overpopulation problem.

Another approach that may alleviate some of the difficulty is the Minor Use Minor Species Animal Health Act, known as the MUMS bill, which proposes to create a program similar to the FDA's human orphan drug program, to facilitate the approval of drugs for use in domestic species such as sheep and goats as well as free-ranging and captive wildlife. The MUMS bill was drafted by a coalition of animal health groups following the 1996 passage of the Animal Drug Availability Act. Unfortunately, although the bill was introduced in Congress in both 2001 and 2002, it was not passed (Bren 2002).

Most new methods being tested for animals are vaccine based, despite the improbability that vaccines would be approved under current FDA policy. Further, the potential for deleterious effects associated with androgen-based methods may well make them inappropriate for wildlife. Even the matrix of the deslorelin implant does not meet current FDA requirements, preventing its approval for marketing in the United States. In fact, most of the methods described here stand little chance of FDA approval unless the guidelines for animal drugs are amended. The result is that the choices faced by zoo and wildlife managers are at best limited and at worst unsafe.

ANTICIPATING FUTURE NEEDS

The need for contraception continues to expand across taxonomic groups, as more species fall under the umbrella of captive breeding programs and as reproductive and survival rates increase with improvements in nutrition, veterinary care, and husbandry. Burgeoning populations of free-ranging wildlife unchecked by predators or hunters may in some cases be amenable to contraception or sterilization. Also, as human encroachment fragments the landscape, carving up wildlife habitat into smaller and smaller segments, wildlife reserves will come to resemble zoos, causing wildlife management to approach captive management. Limited space will constrain the carrying capacity, requiring population sizes to be regulated. Eventually, there may be a need to restrict reproduction in almost all mammalian species, either as part of managed breeding or as population control programs. Species differences will present more challenges to providing safe and effective methods.

Although bird reproduction is primarily controlled by either separation of males from females or by removing eggs, there are some clinical indications for progestin contraception, such as threats to a female's health from the nutrient demands of egg production or the agitation or aggression related to the reproductive cycle. Depo-Provera was used successfully in a variety of parrots (Harrison 1989), suggesting that progestins can also be effective in avian species. Other taxa may also become candidates for contraceptive control of reproductive processes.

The major challenges for free-ranging wildlife remain delivery and species specificity. It is encouraging that there are active research programs for contraceptive testing and development in several countries, as the attendance at recent international programs showed. More ideas and more trials can only help, but differences in regulatory issues and public attitudes may limit application of some approaches. For example, the viral and bacterial products under development in Australia (Bradley 1997) are not likely to be acceptable in the United States. Similarly, the variety of target species may also limit applicability of new methods, for example, those for elephants in Africa compared to those for rabbits in Australia or coyotes in the United States.

The AZA Wildlife Contraception Center has relied primarily on retrospective surveys of contraceptive use in zoos for assessing efficacy and reversibility. This approach has been successful in that it has provided the data that inform the recommendations we provide to the zoo community. However, despite more than a decade of monitoring, there are frustratingly few complete data sets, that is, those that include sufficient repetitions for even the most common methods per species to be confident that the proper dosages are being used. The number of attempts at reversal are even fewer, which severely limits our ability to assess the long-term effects on reproductive potential of animals that have been contracepted. When considering the number of species that currently require contraception for responsible management, the prospects of conducting systematic dose–response and reversal studies quickly become overwhelming. Nevertheless, identifying a manageable number of representative species for prospective study could prove invaluable. The first step might be a complete review of the status of and needs for contraception by taxon.

FUTURE OF CONTRACEPTIVE MONITORING

Through the Contraception Database, the AZA Contraception Advisory Group has monitored contraceptive use in zoo mammals for more than 13 years. The database was originally developed for AZA zoos, but its scope has expanded to include numerous international zoos as well. Data collection has been accomplished through paper-based surveys, but discussions were underway in the late 1990s to develop a software program that would allow Web-based surveys. At the same time, other zoo professionals were identifying the need for an improved, interconnected, and more complete animal records system. The discussions and the multiple needs identified by numerous stakeholders within the global zoological community led to the Zoological Information Management System (ZIMS) project. In brief, the goal of the ZIMS project is to develop and maintain a Web-enabled comprehensive records system that supports the inventory, veterinary, husbandry, management, and conservation work carried out in today's zoos and aquariums. The ZIMS project has the potential to significantly enlarge the breadth and value of the Contraception Database by increasing the number of institutions that contribute data, improving data accuracy (through immediate and single-entry data), and expanding data analysis through enhanced querying and correlating capabilities.

IMPROVING ANIMAL WELFARE

The ethical discussions that surround the use of contraception often revolve around our perceptions or definitions of animal welfare and animal well-being. Although we will likely never be able to reach into the mind of another species, much as we cannot enter the mind of other humans, our ability to evaluate animal well-being is certain to improve in the future. The interest in and research directed at developing metrics to measure the psychological well-being of captive wildlife is greater now than it has ever been. Future research should include developing methods by which the short- and long-term implications of contraception on the social lives of captive and wild animals can be evaluated. As results from empirical evaluations of animal well-being become available, they should be incorporated into our animal management programs. The information gained should help managers select methods and management protocols that best meet animal welfare concerns while addressing the need to limit reproduction in captive and free-ranging wildlife.

The AZA Wildlife Contraception Center, along with other organizations interested in the application of contraceptives to the management of captive and free-ranging wildlife, will continue to pursue research and development of promising new techniques, to advocate for their approval by the FDA, and to promote their use.

ACKNOWLEDGMENTS

The authors thank Drs. L. J. D. Zaneveld and Wolfgang Jöchle for suggestions on the manuscript.

REFERENCES

Anderson, R. A., H. Zhu, L. Cheng, and D. T. Baird. 2002. Investigation of a novel preparation of testosterone decanoate in men: pharmacokinetics and spermatogenic suppression with etonogestrel implants. *Contraception* 66:357–364.

Asa, C. S., and I. Porton. 1991. Concerns and prospects for contraception in carnivores. In *Proceedings, American Association of Zoo Veterinarians*, 298–303, September 28–October 3, 1991, Calgary, Alberta, Canada.

Asa, C. S., L. J. D. Zaneveld, L. Munson, M. Callahan, and A. P. Byers. 1996. Efficacy, safety and reversibility of a bisdiamine as a male-directed oral contraceptive in gray wolves (*Canis lupus*). *J Zoo Wildl Med* 27:501–506.

Beyler, A. L., G. O. Poots, F. Coulston, and A. R. Surrey. 1961. The selective testicular effects of certain bis(dichloroacetyl)diamines. *Endocrinology* 69:819–833.

Bradley, M. P. 1997. Immunocontraceptive vaccines for control of fertility in the European red fox (*Vulpes vulpes*). In *Contraception in Wildlife Management*, ed. T. J. Kreeger, 195–203. Technical Bulletin No. 1853. Washington, DC: US Dept. of Agriculture, Animal and Plant Health Inspection Service.

Bren, L. 2002. Treating minor species: a major animal health concern. *FDA Vet* 17:7–12.

Chaki, S. P., H. C. Das, and M. M. Misro. 2003. A short-term evaluation of semen and accessory gland function in phase III trial subjects receiving intravasal contraceptive RISUG. *Contraception* 67:73–78.

Cheng, C. Y., M. Mo, J. Grima, L. Saso, B. Tita, D. Mruk, and B. Silvestrini. 2002. Indazole carboxylic acids in male contraception. *Contraception* 65:265–268.

Coe, C. L., A. Savage, and L. J. Bromley. 1992. Phylogenetic influences on hormone levels across the primate order. *Am J Primatol* 28:81–100.

Coonrod, S. 2002. Identification of new contraceptive targets in the oocyte. In *Proceedings, 2002 International Symposium on Nonsurgical Methods of Pet Population Control*, 54–55, April 19–21, 2002, Pine Mountain, GA. Alliance for Contraception in Cats and Dogs.

Coutinho, E. M. 2002. Gossypol: a contraceptive for men. *Contraception* 65:259–263.

DeMatteo, K. E., I. J. Porton, and C. S. Asa. 2002. Comments from the AZA Contraception Advisory Group on evaluating the suitability of contraceptive methods in golden-headed lion tamarins (*Leontopithecus chrysomelas*). *Anim Welf* 11:343–348.

De Vleeschouwer, K., K. Leus, and L. Van Elsacker. 2000. An evaluation of the suitability of contraceptive methods in golden-headed lion tamarins (*Leontopithecus chrysomelas*), with emphasis on melengestrol acetate (MGA) implants. (I) Effectiveness, reversibility and medical side-effects. *Anim Welf* 9:251–271.

Drobeck, H. P., and F. Coulston 1962. Inhibition and recovery of spermatogenesis in rats, monkeys, and dogs medicated with bis(dichloroacetyl)-diamines. *Exp Mol Pathol* 1:251–274.

Fail, P. A., S. A. Anderson, and C. E. Cook. 2000. 28-day toxicology test: indenopyridine RTI 4587-056 in male Sprague-Dawley rats. *Reprod Toxicol* 14:265–274.

Farcon, E., R. S. Hotchkiss, and E. S. Nuwayser. 1975. An absorbable intravasal stent and a silicone intravasal plug. *Invest Urol* 13:108–112.

Gatto, M. T., B. Tita, M. Artico, and L. Saso. 2002. Recent studies on lonidamine, the lead compound of the antispermatogenic indazol-carboxylic acids. *Contraception* 65:277–278.

Gupta, I., S. Dhawan, G. D. Goel, and K. Saha. 1975. Low fertility rate in vasovasostomized males and its possible immunologic mechanism. *Int J Fertil* 20:183–191.

Hardy, C. M., H. G. Clarke, B. Nixon, J. A. Grigg, L. A. Hinds, and M. K. Holland. 1997. Examination of the immunocontraceptive potential of recombinant rabbit fertilin subunits in rabbit. *Biol Reprod* 57:879–886.

Harrison, G. J. 1989. Medroxyprogesterone acetate-impregnated silicone implants: preliminary use in pet birds. *Proc Annu Conf Assoc Avian Vet* 1989:6–10.

Heller, C. G., D. J. Moore, and C. A. Paulsen. 1961. Suppression of spermatogenesis and chronic toxicity in man of a new series of bis(dichloroacetyl)diamines. *Toxicol Appl Pharmacol* 3:1–11.

Hodel, C., and K. Suter. 1978. Reversible inhibition of spermatogenesis with an indenopyridine (20-438). *Arch Toxicol* (Suppl) 1:323–326.

Kibble, M. W., and M. B. Ross. 1987. Adversive effects of anabolic steroids in athletes. *Clin Pharmacol* 6:686–692.

Lea, I. A., M. J. van Lierop, E. E. Widgren, A. Grootenhuis, Y. Wen, M. van Duin, and M. G. O'Rand. 1998. A chimeric sperm peptide induces antibodies and strain-specific reversible infertility in mice. *Biol Reprod* 59:527–536.

Lee, H. Y. 1969. Experimental studies on reversible vas occlusion by intravasal thread. *Fertil Steril* 20:735–744.

Liu, G. Z. 1987. Clinical studies of gossypol as a male contraceptive. *Reproduction* 5:189–192.

Lohiya, N. K., B. Manivannan, and P. K. Mishra. 1998. Ultrastructural changes in the spermatozoa of langur monkeys, *Presbytis entellus entellus*, after vas occlusion with styrene maleic anhydride. *Contraception* 57:125–132.

Malone, D. A., R. J. Dimef, J. A. Lombardo, and R. H. Sample. 1995. Psychiatric effects and psychoactive substance use in anabolic-androgenic steroid users. *Clin J Sports Med* 5:25–31.

Manivannan, B., P. K. Mishra, and N. K. Lohuja. 1999. Ultrastructural changes in the vas deferens of langur monkeys, *Presbytis entellus entellus*, after vas occlusion with styrene maleic anhydride and after its reversal. *Contraception* 59:137–144.

Massai, M. R., S. Diaz, E. Quinteros, M. V. Reyes, C. Herreros, A. Zepeda, H. B. Croxatto, and A. J. Moo-Young. 2001. Contraceptive efficacy and clinical performance of Nestorone implants in postpartum women. *Contraception* 64:369–376.

Matter, B. E., I. Jaeger, W. Sutet, T. Tsuchimoto, and D. Deyssenroth. 1979. Actions of an anti-spermatogenic, but non-mutagenic, indenopyridine derivative in mice and *Salmonella typhinurium*. *Mutat Res* 66:113–127.

Mazumdar, S., and A. S. Levine. 1998. Antisperm antibodies: etiology, pathogenesis, diagnosis and treatment. *Fertil Steril* 70:799–810.

McLachlin, R. I., J. McDonald, D. Rushford, D. M. Robertson, C. Garrett, and H. W. G. Baker. 2000. Efficacy and acceptability of testosterone implants, alone or in combination with a 5α-reductase inhibitor, for male contraception. *Contraception* 62:73–78.

Meriggiola, M. C., W. J. Bremmer, A. Costantino, G. Di Cintio, and C. Flamigni. 1998. Low dose of cyproterone acetate and testosterone enanthate for contraception in men. *Hum Reprod* 13:1225–1229.

Meriggiola, M. C., A. Costantino, and S. Cerpolini. 2002. Recent advances in hormonal male contraception. *Contraception* 65:269–272.

Metz, C., K. Hohl, S. Waidelich, W. Drochner, and R. Claus. 2002. Active immunization of boars against GnRH at an early age: consequences for testicular function, boar taint accumulation and N-retention. *Livest Prod Sci* 74:147–157.

Middleman, A. B., and R. H. DuRant. 1996. Anabolic steroid use and associated health risk behaviours. *Sports Med* 21:221–225.

Mohr, K. L., and P. T. Johnson. 1978. Vasovasostomy and vas occlusion: preliminary observations using artificial devices in guinea pigs. *Fertil Steril* 30:696–701.

Naz, R. K., and S. C. Chauhan. 2002. Human sperm-specific peptide vaccine that causes long-term reversible contraception. *Biol Reprod* 67:674–680.

Plotka, E. D., T. C. Eagle, D. N. Vevea, A. L. Koller, D. B. Siniff, J. R. Tester, and U. S. Seal. 1988. Effects of hormone implants on estrus and ovulation in feral mares. *J Wildl Dis* 24:507–514.

Primakoff, P., W. Lathrop, L. Wollman, A. Cowan, and D. Myles. 1988. Fully effective contraception in male and female guinea pigs immunized with the sperm protein PH-20. *Nature* (Lond) 335:543–547.

Ramachandra, S. G., V. Ramesh, H. N. Krishnamurthy, N. Kumar, K. Sundarum, M. P. Hardy, and A. J. Rao. 2002. Effect of chronic administration of 7α-methyl-19-nortestosterone on serum testosterone, number of spermatozoa and fertility in adult male bonnet monkeys (*Macaca radiata*). *Reproduction* 124:301–309.

Robbins, S. C. 2002. Active immunization of prepubertal cats against gonadotropin releasing hormone and its effects on gonadal hormone concentration and reproductive function. In *Proceedings, 2002 International Symposium on Nonsurgical Methods of Pet Population Control*, 51–53, April 19–21, 2002, Pine Mountain, GA. Alliance for Contraception in Cats and Dogs.

Silber, S. J. 1979. Epididymal extravasation following vasectomy as a cause for failure of vasectomy reversal. *Fertil Steril* 31:309–315.

Wang, M. 2002. Neutersol: intratesticular injection induces sterility in dogs. In *Proceedings, 2002 International Symposium on Nonsurgical Methods of Pet Population Control*, 62–65, April 19–21, 2002, Pine Mountain, GA. Alliance for Contraception in Cats and Dogs.

Webel, S. K., and E. L. Squires. 1982. Control of the oestrous cycle in mares with altrenogest. *J Reprod Fertil Suppl* 32:193–198.

Weinbauer, G. F., E. Rovan, and J. Frick. 1983. Toxicity of gossypol at antifertility dosages in male rats. Statistical analyses of lethal rates and body weight responses. *Andrologia* 15:213–221.

World Health Organization (WHO) Task Force on Methods for the Regulation of Male Fertility. 1990. Contraceptive efficacy of testosterone-induced azoospermia in normal men. *Lancet* 336:955–959.

World Health Organization (WHO) Task Force. 1994a. Progestogen-only contraceptives during lactation. I. Infant growth. *Contraception* 50:35–54.

World Health Organization (WHO) Task Force. 1994b. Progestogen-only contraceptives during lactation. II. Infant development. *Contraception* 50:55–68.

World Health Organization (WHO) Task Force on Methods for the Regulation of Male Fertility. 1996. Contraceptive efficacy of testosterone-induced azoospermia and oligospermia in normal men. *Fertil Steril* 65:821–829.

Zaneveld, L. J. D., J. W. Burns, S. Beyler, W. Depel, and S. Shapiro. 1988. Development of a potentially reversible vas deferens occlusion device and evaluation in primates. *Fertil Steril* 49:527–533.

Zhu, X., and R. K. Naz 1997. Fertilization antigen-1: cDNA cloning, testis-specific expression and immunocontraceptive effects. *Proc Natl Acad Sci USA* 94:4704–4709.

APPENDIX

Total number of individuals, by species, represented in the American Zoo and Aquarium Association Contraception Advisory Group (AZA CAG) Contraception Database and the contraceptive methods used.

Order/family	Genus/species	Common name	♀/♂	Method
Carnivora				
Felidae	*Acinonyx jubatus*	Cheetah	7/0	MGA
	Caracal caracal	Caracal	10/1	MGA, VP
	Felis chaus	Jungle cat	5/0	MGA
	Felis margarita	Sand cat	2/0	MGA
	Felis nigripes	Black-footed cat	1/0	MGA
	Felis silvestris	Wild cat	9/2	MGA, CT
	Herpailurus yagouaroundi	Jaguarundi	3/0	MGA
	Leopardus pardalis	Ocelot	12/0	MGA, PRp
	Leopardus wiedii	Margay	3/1	MGA, CT
	Leptailurus serval	Serval	23/1	MGA, VS
	Lynx canadensis	Canadian lynx	11/1	MGA, MA, VS
	Lynx lynx	Northern lynx	18/1	MGA, PLG, VS
	Lynx rufus	Bobcat	24/8	MGA, MA, VP, VS, CT
	Neofelis nebulosa	Clouded leopard	15/0	MGA, HIS, LUP
	Oncifelis geoffroyi	Geoffroy's cat	2/1	MGA, CT
	Panthera leo	Lion	400/19	MGA, MPA, MA, PRp, LUP, GST, VP, VS, CT
	Panthera onca	Jaguar	89/2	MGA, MPA, MA, PLG, VS, CT

Continued on next page

Order/family	Genus/species	Common name	♀/♂	Method
	Panthera pardus	Leopard	115/5	MGA, MPA, MA, VS, CT
	Panthera tigris	Tiger	314/9	MGA, MPA, MA, PLG, GST, VS, CT
	Prionailurus bengalensis	Leopard cat	3/0	MGA
	Puma concolor	Puma or mountain lion	114/6	MGA, MPA, PRp, VS, CT
	Uncia uncia	Snow leopard	94/2	MGA, MPA, VP, VS
Canidae	*Alopex lagopus*	Arctic fox	2/6	MGA, VS, CT
	Canis latrans	Coyote	0/1	CT
	Canis lupus	Wolf	31/20	MGA, MA, VS, CT
	Canis mesomelas	Black-backed jackal	4/1	MGA, VS
	Canis rufus	Red wolf	46/6	MGA, DSL, HIS
	Chrysocyon brachyurus	Maned wolf	9/0	MGA
	Lycaon pictus	African wild dog	28/0	MGA, MPA
	Nyctereutes procyonoides	Raccoon dog	1/0	MGA
	Otocyon megalotis	Bat-eared dog	3/0	MGA
	Speothos venaticus	Bush dog	2/0	MGA, DSL
	Urocyon cinereoargenteus	Grey fox	0/3	VS, CT
	Vulpes corsac	Corsac fox	1/0	MPA
	Vulpes velox	Swift fox	4/0	MGA
	Vulpes vulpes	Red fox	0/3	VS
	Vulpes zerda	Fennec fox	11/0	MGA, MA, DSL
Ursidae	*Helarctos malayanus*	Malayan sun bear	6/0	MGA, MGAf
	Melursus ursinus	Sloth bear	8/0	MGA
	Tremarctos ornatus	Spectacled bear	19/0	MGA, MPA
	Ursus americanus	North American black bear	23/3	MGA, VS, CT
	Ursus arctos	Brown bear	59/3	MGA, MPA, MA, VS, CT
	Ursus maritimus	Polar bear	30/1	MGA, MPA, MA, VP
	Ursus thibetanus	Asiatic black bear	17/1	MGA, MPA, CT
Procyonidae	*Nasua narica*	White-nosed coati	2/2	MGA, VS, CT
	Nasua nasua	Ringtailed coati	10/0	MGA
	Potos flavus	Kinkajou	2/0	MGA
	Procyon lotor	Raccoon	7/3	MGA, CT
Mustelidae	*Amblonyx cinereus*	Oriental small-clawed otter	20/0	MGA
	Aonyx capensis	African clawless otter	1/0	MGA

Order/family	Genus/species	Common name	♀/♂	Method
	Eira barbara	Tayra	1/0	MGA
	Enhydra lutris	Sea otter	1/0	MGA
	Ictonyx striatus	Zorilla	1/0	MGA
	Lutra canadensis	North American river otter	4/2	MGA, MPA, LUP
	Lutra maculicollis	Spot-necked otter	1/0	MGA
	Mephitis mephitis	Striped skunk	0/1	VS
	Mustela putorius	European polecat	0/4	CT
Viverridae	*Arctictis binturong*	Binturong	10/0	MGA
	Genetta genetta	Small spotted genet	2/0	MGA
	Helogale parvula	Dwarf mongoose	1/1	MGA
	Herpestes sp.	Common mongoose	0/1	CT
	Paguma larvata	Masked palm civet	1/0	MGA
	Suricata suricatta	Meerkat	9/0	MGA, MPA
Hyaenidae	*Crocuta crocuta*	Spotted hyena	3/0	MGA
	Hyaena hyaena	Striped hyena	4/0	MGA
	Proteles cristatus	Aardwolf	5/0	MGA
Pinnipedia				
Otariidae	*Callorhinus ursinus*	Northern fur seal	1/0	MGA, PZP
	Zalophus californianus	California sea lion	21/12	MGA, DP, LUP, PZP, VS, CT
Phocidae	*Halichoerus grypus*	Grey seal	2/0	MGA
	Phoca vitulina	Harbor seal	14/2	MGA, MPA, MA, SUP, LUP
Cetacea				
Delphinidae	*Lagenorhynchus obliquidens*	Pacific white-sided dolphin	2/0	ALT
	Tursiops truncatus	Bottlenose dolphin	5/4	LUP, ALT
Primates				
Lorisidae	*Nycticebus pygmaeus*	Pygmy loris	5/0	MGA, MPA
	Nycticebus coucang	Slow loris	4/0	MGA, MPA
	Perodicticus potto	Potto	2/0	MPA
	Otolemur crassicaudatus	Greater bush baby	12/0	MGA, MPA
Lemuridae	*Hapalemur griseus*	Gentle grey lemur	16/0	MGA, MPA
	Eulemur coronatus	Crowned lemur	10/0	MGA, MPA
	Eulemur mongoz	Mongoose lemur	25/0	MGA, MPA
	Eulemur rubriventer	Red-bellied lemur	9/0	MGA, MPA
	Eulemur macaco	Black lemur	70/6	MGA, MPA, DSL
	Eulemur fulvus	Brown lemur	69/0	MGA, MPA

Continued on next page

Order/family	Genus/species	Common name	♀/♂	Method
	Lemur catta	Ringtailed lemur	86/6	MGA, MPA, DSL, VP
	Varecia variegata	Ruffed lemur	214/1	MGA, MPA, LNG, DSL, VS
Indriidae	*Propithecus verreauxi*	Coquerel's sifaka	1/0	MGA
Callitrichidae	*Callithrix pygmaea*	Pygmy marmoset	24/0	MGA, MPA
	Callithrix jacchus	Common marmoset	26/0	MGA
	Callithrix penicillata	Black-eared marmoset	1/0	MPA
	Callithrix geoffroyi	Geoffroy's marmoset	14/0	MGA
	Callithrix kuhlii	Black tufted-ear marmoset	3/0	MGA
	Saguinus midas	Red-handed tamarin	8/0	MGA, MPA
	Saguinus labiatus	White-lipped tamarin	1/0	MGA
	Saguinus imperator	Emperor tamarin	8/0	MGA, MPA
	Saguinus fuscicollis	Saddle-backed tamarin	2/0	MGA
	Saguinus geoffroyi	Geoffroy's tamarin	1/0	MGA, MPA
	Saguinus oedipus	Cotton-topped tamarin	208/1	MGA, MPA, LNG, VP
	Leontopithecus chrysopygus	Black lion tamarin	1/0	MGA
	Leontopithecus rosalia	Golden lion tamarin	162/1	MGA, MPA
	Leontopithecus chrysomelas	Golden-headed lion tamarin	54/0	MGA, MPA
	Callimico goeldii	Goeldi's monkey	63/5	MGA, MPA, DSL
Cebidae	*Saimiri sciureus*	Squirrel monkey	23/2	MGA, MPA
	Cebus apella	Brown capuchin	12/0	MGA
	Cebus capucinus	White-throated capuchin	5/0	MGA, MPA
	Cebus olivaceus	Weeper capuchin	1/0	MGA
	Cebus albifrons	Brown pale-fronted capuchin	1/0	MGA
	Aotus lemurinus	Grey-legged douroucouli	2/0	MGA
	Aotus azarai	Douroucouli	1/0	MGA
	Aotus vociferans	Noisy douroucouli	1/0	MGA
	Aotus trivirgatus	Douroucouli	9/0	MGA
	Callicebus torquatus	Titi monkey	1/0	MGA

Order/family	Genus/species	Common name	♀/♂	Method
	Callicebus moloch	Dusky titi	6/0	MGA
	Callicebus donacophilus	Reed titi	7/0	MGA, MPA
	Pithecia pithecia	White-faced saki	30/0	MGA, MPA, LNG, MGA/EE
	Alouatta caraya	Black howler monkey	33/1	MGA, MPA, VP
	Alouatta palliata	Mantled howler	6/0	MGA, MPA
	Ateles fusciceps	Brown-headed spider monkey	12/1	MGA, MPA
	Ateles geoffroyi	Black-handed spider monkey	37/1	MGA, MPA, LUP, MGA/EE
Cercopithecidae	*Erythrocebus patas*	Patas monkey	41/3	MGA, MPA, LNG
	Chlorocebus aethiops	Vervet	5/1	MGA
	Allenopithecus nigroviridis	Allen's swamp monkey	2/0	MGA, MPA
	Cercopithecus wolfi	Wolf's guenon	1/0	MGA
	Cercopithecus ascanius	Guenon	12/0	MGA, MPA
	Cercopithecus diana	Diana monkey	3/0	MGA
	Cercopithecus cephus	Moustached monkey	1/0	MGA, MPA
	Cercopithecus mitis	Sykes' monkey	12/0	MGA
	Cercopithecus mona	Mona monkey	7/0	MGA
	Cercopithecus neglectus	DeBrazza's monkey	20/0	MGA, LNG, ETN
	Cercocebus galeritus	Agile mangabey	3/0	MGA
	Cercocebus torquatus	Sooty mangabey	20/0	MGA, MPA
	Cercocebus agilis	Agile mangabey	7/0	MGA
	Lophocebus albigena	White-cheeked mangabey	12/0	MGA, MPA
	Macaca silenus	Lion-tailed macaque	53/8	MGA, MPA, LNG, HIS, DSL,
	Macaca nigra	Sulawesi macaque	11/0	MGA, LNG
	Macaca nemestrina	Pig-tailed macaque	4/0	MGA, LNG
	Macaca fuscata	Japanese macaque	74/2	MGA, MPA, VP
	Macaca fascicularis	Crab-eating macaque	3/1	MGA, VP
	Macaca maura	Moor macaque	7/0	MGA
	Macaca mulatta	Rhesus macaque	51/0	MGA
	Macaca sylvanus	Barbary macaque	12/3	MGA
	Macaca tonkeana	Tonkean macaque	6/0	MGA

Continued on next page

Order/family	Genus/species	Common name	♀/♂	Method
	Papio hamadryas	Hamadryas baboon	180/5	MGA, MPA, LNG, VP
	Mandrillus sphinx	Mandrill	54/1	MGA, MPA, LNG, MA, BCp, VP
	Theropithecus gelada	Gelada	1/0	MPA
	Colobus polykomos	Western black and white colobus	1/0	MGA
	Colobus guereza	Black and white colobus	102/0	MGA, MPA, LNG, ETN
	Trachypithecus cristatus	Silvered leaf langur	8/0	MGA, LNG
	Trachypithecus francoisi	Francois' langur	18/0	MGA, MPA, LNG
	Trachypithecus obscurus	Dusky-leaf monkey	19/0	MGA, MPA
	Semnopithecus entellus	Hanaman langur	7/0	MGA, MPA
	Pygathrix nemaeus	Douc langur	4/0	MGA, MPA
Hylobatidae	*Hylobates lar*	Lar gibbon	65/2	MGA, MPA, LNG, BCp, VP
	Hylobates leucogenys	White-cheeked gibbon	4/0	MGA, PRp
	Hylobates agilis	Agile gibbon	1/0	MGA
	Hylobates concolor	Crested gibbon	1/0	MGA
	Hylobates syndactylus	Siamang	54/1	MGA, MPA, LNG, MA, BCp, VP
Pongidae	*Pongo pygmaeus*	Orangutan	132/7	MGA, MPA, LNG, ETN, MGA/EE, BCp, VP, VS, IUD
	Pan troglodytes	Chimpanzee	161/4	MGA, MPA, LNG, ETN, BCp, PRp, VP
	Gorilla gorilla	Gorilla	39/0	MGA, BCp
Artiodactyla				
Suidae	*Babyrousa babyrussa*	Babirusa	2/0	PZP
	Phacochoerus aethiopus	Desert warthog	2/0	MGA, MPA
	Phacochoerus africanus	Warthog	3/0	MPA
	Sus scrofa	Wild boar	1/0	MGA
Tayassuidae	*Pecari tajacu*	Collared peccary	1/0	MGA
Hippopotamidae	*Hippopotamus amphibius*	Hippopotamus	19/0	MGA, MPA, MA, LUP, PZP
Camelidae	*Camelus bactrianus*	Bactrian camel	10/1	MGA, MPA, LUP, PZP
	Camelus dromedarius	Dromedary camel	5/0	MGA
	Lama glama	Llama	5/0	MGA
	Lama guanicoe	Guanaco	5/0	MGA, MPA
	Lama pacos	Alpaca	1/1	MPA, CT

Order/family	Genus/species	Common name	♀/♂	Method
Tragulidae	*Tragulus napu*	Larger mouse deer	1/0	MGA
Cervidae	*Alces alces*	Moose	3/0	MGA, MPA
	Axis axis	Spotted deer	0/1	MGA
	Cervus albirostris	White-lipped deer	4/0	MGA
	Cervus duvauceli	Barasingha	3/0	MGA, PZP
	Cervus elaphus	Elk	10/0	MGA, MPA
	Cervus eldii	Eld's deer	5/0	MGA, PZP
	Cervus nippon	Sika deer	6/0	MGA
	Dama dama	Fallow deer	0/3	VP
	Elaphurus davidianus	Pere David's deer	1/0	PZP
	Odocoileus hemioneus	Mule deer	3/2	MGA, MPA
	Odocoileus virginianeus	White-tailed deer	17/2	MGA, VS
	Muntiacus reevesi	Reeve's muntjac	15/1	MGA, MPA, PZP
	Rangifer tarandus	Reindeer	0/3	MPA
Giraffidae	*Giraffa camelopardalis*	Giraffe	31/1	MGA, MPA, LUP, PZP
Bovidae	*Aeryceros melampus*	Impala	2/0	PZP
	Antidorcas marsupialis	Springbok	2/0	MGA, PZP
	Antilope cervicapra	Blackbuck	1/0	MGA
	Addax nasomaculatus	Addax	20/4	MGA, PZP
	Bison bison	American bison	8/0	MPA, PZP
	Bison bonasus	European bison	4/0	MGA, DSL
	Bos frontalis	Gaur	5/4	MGA, LUP
	Bos grunniens	Yak	1/0	MGA
	Bos javanicus	Banteng	5/0	MGA, MPA, PZP
	Bos taurus	Cow/ox	1/0	MGA
	Boselaphus tragocamelus	Nilgai	5/1	MGA
	Bubalus depressicornis	Lowland anoa	2/0	MGA, PZP
	Capra falconeri	Markhor	6/0	LNG, PZP
	Capra hircus	Goat	1/0	PZP
	Capra ibex	Ibex	10/0	MGA, PZP
	Capra nubiana	Nubian ibex	3/0	MGA
	Capra sibirica	Siberian ibex	9/0	MGA, PZP
	Cephalophus maxwellii	Maxwell's duiker	1/0	MGA
	Cephalophus monticola	Blue duiker	1/0	MGA
	Cephalophus rufilatus	Red-flanked duiker	1/0	MPA
	Damaliscus lunatus	Topi	4/0	MGA, PZP
	Damaliscus pygargus	Blesbok	2/0	MGA
	Gazella cuveiri	Cuvier gazelle	5/0	MGA, PZP
	Gazella dama	Dama gazelle	6/0	MGA, PZP

Continued on next page

Order/family	Genus/species	Common name	♀/♂	Method
	Gazella spekei	Speke's gazelle	2/0	MGA
	Gazella thomsoni	Thompson's gazelle	2/0	MGA
	Hemitragus hylocrius	Nilgiri tahr	11/3	MGA, PZP
	Hemitragus jemlahicus	Himalayan tahr	9/0	MGA, MPA, PZP
	Hippotragus niger	Sable antelope	9/0	MGA, MPA, PZP
	Kobus ellipsiprymnus	Common waterbuck	4/0	PZP
	Kobus megaceros	Nile lechwe	8/0	MGA, DSL
	Madoqua guentheri	Gunther's dik dik	1/5	MGA, VP, VS
	Madoqua kirki	Kirk's dik dik	7/0	MGA, PZP
	Naemorhedus caudatus	Central Chinese goral	3/0	MGA
	Naemorhedus goral	Common goral	13/0	MGAf
	Oreamnos americanus	Rocky mountain goat	4/0	MGA, LNG, PZP
	Oreotragus oreotragus	Klipspringer	1/0	MGA
	Oryx dammah	Scimitar-horned oryx	0/12	MGA, GnRH
	Oryx leucoryx	Arabian oryx	18/0	MGA, PZP
	Ovibos moschatus	Musk oxen	4/0	MPA
	Ovis canadensis	Bighorn sheep	11/1	MGA, MPA, LUP
	Ovis dalli	Dall's sheep	6/0	MGA, MPA
	Ovis vignei	Arkal urial sheep	8/0	MGA
	Rupicapra rupicapra	Chamois	9/0	MGA, PZP
	Syncerus caffer	African buffalo	2/0	MGA
	Taurotragus oryx	Common eland	6/0	MGA
	Tragelaphus angasi	Nyala	1/0	PZP
	Tragelaphus eurycerus	Bongo	10/0	MGA, MPA, DSL, PZP
	Tragelaphus scriptus	Bushbuck	1/0	MPA
	Tragelaphus spekei	Sitatunga	4/0	MGA, DSL, PZP
	Tragelaphus strepsiceros	Greater kudu	3/2	MGA, MPA, PZP
Perissodactyla				
Equidae	*Equus asinus*	Domestic ass	1/0	PZP
	Equus burchelli	Common zebra	7/0	PZP
	Equus grevyi	Grevy's zebra	15/0	PZP
	Equus zebra	Hartman's mountain zebra	2/0	PZP
Tapiridae	*Tapirus bairdii*	Baird's tapir	1/0	MGA, MPA
	Tapirus indicus	Malayan tapir	1/0	MGA, MPA
	Tapirus terrestris	Brazilian tapir	9/0	MGA, MPA, PZP

Order/family	Genus/species	Common name	♀/♂	Method
Proboscidea				
Elephantidae	*Elephas maximus*	Asian elephant	1/0	PZP
Hyracoidea				
Procaviidae	*Procavia capensis*	Rock hyrax	5/0	MGA
Scandentia				
Tupaiidae	*Tupaia glis*	Common tree shrew	1/0	MGA
Rodentia				
Castoridae	*Castor canadensis*	American beaver	3/0	MGA, LNG
Pedetidae	*Pedetes capensis*	Springhaas	1/0	MGA
Muridae	*Hypogeomys antimena*	Malagasy giant jumping rat	2/0	LNG
Erethizontidae	*Coendou prehensilis*	Prehensile-tailed porcupine	1/0	MGA
Caviidae	*Dolichotis patagonum*	Patagonian cavy or mara	10/1	MGA, LNG, DSL
Hydrochoeridae	*Hydrochaeris hydrochaeris*	Capybara	8/0	MGA
Myocastoridae	*Myocastor coypus*	Coypu	1/0	MGA
Capromyidae	*Capromys* sp.	Hutia	1/0	MGA
Hystricidae	*Hystrix cristata*	Crested porcupine	1/0	MGA
Xenarthra				
Myrmecophagidae	*Tamandua tetradactyla*	Southern tamandua	1/0	MGA
Megalonychidae	*Choloepus hoffmanni*	Hoffman's two-toed sloth	1/0	MGA
Chiroptera				
Pteropodidae	*Pteropus rodricensis*	Rodrigues flying fox	31/0	MGA
Marsupialia				
Macropodidae	*Dendrolagus bennettianus*	Bennett's tree kangaroo	1/0	MGA
	Dendrolagus matschiei	Matschie's tree kangaroo	1/0	MGA
	Macropus bernardus	Black wallaroo	1/0	LNG
	Macropus fuliginosus	Western grey kangaroo	4/0	LNG
	Macropus giganteus	Great grey kangaroo	2/0	MGA, LNG
	Macropus robustus	Common wallaroo	5/0	MGA, MPA
	Macropus rufus	Red kangaroo	2/3	MGA, VP

Continued on next page

Order/family	Genus/species	Common name	♀/♂	Method
	Petrogale xanthopus	Yellow-footed rock wallaby	1/1	LNG, VS
	Thylogale stigmatica	Red-legged pademelon	2/0	LNG
	Wallabia bicolor	Swamp wallaby	1/0	LNG
Phalangeridae	*Strigocuscus gymnotis*	Ground cuscus	1/0	MGA

ALT, altrenogest (Regu-mate); BCp, estrogen–progestin combination birth control pill; CT, castration; DSL, deslorelin GnRH agonist implant; ETN, etonorgestrel implant (Implanon); GST, gestapuran; GnRH, gonadotropin-releasing hormone agonist (generic); HIS, histrelin GnRH agonist implant; IUD, intrauterine device; LNG, levonorgestrel implant (Norplant or Jadelle); LUP, Lupron Depot GnRH agonist injection; MA, megestrol acetate pill (Ovaban, Megace, Ovarid); MGA, melengestrol acetate implant; MGAf, MGA in feed; MGA/EE, MGA plus ethinyl estradiol implant; MPA, medroxyprogesterone acetate injection (Depo-Provera, Provera); PLG, proligestone; PRp, progestin-only pill (generic); PZP, porcine zona pellucida vaccine; SUP, Supprestral; VP, vas plug; VS, vasectomy.

CONTRIBUTORS

Cheryl S. Asa, PhD
Co-Director
AZA Wildlife Contraception Center
Director of Research
Saint Louis Zoo
One Government Drive
St. Louis, MO 63110-1395

Paul P. Calle, VMD, DIPL ACZM
Senior Veterinarian
Wildlife Health Sciences
Wildlife Conservation Society
2300 Southern Blvd.
Bronx, NY 10460-1099

Karen E. DeMatteo, PhD
Contraception Database Manager
Saint Louis Zoo
One Government Drive
St. Louis, MO 63110-1395

Kimberly Frank
Science and Conservation Center
ZooMontana
P.O. Box 80905
Billings, MT 59108

Wolfgang Jöchle, DVM
Consulting Theriogenologist
Manahawkin, NJ 08050

Jay Kirkpatrick, PhD
Director, Science and Conservation Center
ZooMontana
P.O. Box 80905
Billings, MT 59108

Anneke Moresco, DVM
University of California
School of Veterinary Medicine
Department of Pathology, Microbiology, and Immunology
Davis, CA 95616

Linda Munson, DVM, PhD
University of California
School of Veterinary Medicine
Department of Pathology, Microbiology, and Immunology
Davis, CA 95616

Marilyn L. Patton, MS
CRES
Zoological Society of San Diego
P.O. Box 120551
San Diego, CA 92112-0551

Linda Penfold, PhD
White Oak Conservation Center
3823 Owens Road
Yulee, FL 32097

Ingrid Porton, MS
Co-Director
AZA Wildlife Contraception Center
Primate Curator
Saint Louis Zoo
One Government Drive
St. Louis, MO 63110-1395

INDEX